Accelerating IoT Development with ChatGPT

A practical guide to building your first IoT project
using AI-assisted coding and cloud integration

Jun Wen

Accelerating IoT Development with ChatGPT

Group Product Manager: Preet Ahuja

Publishing Product Manager: Vidhi Vashisth

Book Project Manager: Ashwin Dinesh Kharwa

Senior Editor: Mudita S

Technical Editor: Arjun Varma

Copy Editor: Safis Editing

Proofreader: Mudita S

Indexer: Tejal Soni

Production Designer: Nilesh Mohite, Joshua Misquitta

DevRel Marketing Coordinator: Rohan Dobhal

First published: August 2024

Production reference: 1020824

Published by Packt Publishing Ltd.

Grosvenor House

11 St Paul's Square

Birmingham

B3 1RB, UK.

ISBN 978-1-83546-162-4

www.packtpub.com

Contributors.

About the author

Jun Wen, the founder of AI Discovery Academy, is a passionate evangelist for AI enlightenment education among school students. With more than twenty years of experience in technology development, Jun has specialized in a wide range of industrial domains, including 4G/LTE, 5G, Wi-Fi, BLE, LoRaWAN, IoT, robotics, and AI. He has previously held senior product management positions at Fortune 500 companies such as Amazon, Cisco, and Motorola. Jun holds a Master of Science degree from Brown University and is an AWS Certified Solutions Architect - Professional. His passion extends to creating IoT innovations, utilizing platforms including Arduino, Raspberry Pi, RISC-V, and the AWS cloud, and finding practical applications for AI.

I want to express heartfelt thanks to my beloved wife, Yajing Wu, and my children, Suixin Wen and Ryan Wen, for their support during the writing of this book. Gratitude also goes to my Brown University's Master's in Technology Leadership Program Cohort 6 friends and professors, whose enthusiasm and insights sparked my journey. Finally, I appreciate Ashwin Dinesh Kharwa, Mudita, and Vidhi Vashisth from the Packt team for their professional service.

About the reviewers

Bob Lo has been working in networking technologies for over 30 years, with 25 years of them spent developing networking software at Cisco Systems. He is highly experienced in network/protocols, L2/L3, switches, routers, ASIC design, network drivers, embedded devices, and more. In the last 10+ years at Cisco, he has led a big team across the globe developing and innovating the latest IoT technologies. He received his BS in computer science from the National University of Singapore and an MSc in computer science from North Dakota State University. He has also worked with researchers and professors from the University of Illinois, Urbana-Champaign, on cutting-edge network technologies.

Sebastián Viviani is an electronics "all-rounder." His passion for technology started at the age of 6, when he got a kids' book about lights and switches, followed by a Commodore 64 and writing his first programs in BASIC.

The learning journey has never stopped since. He graduated as an electronics engineer from U.T.N. (Argentina) and has worked for several industry leaders including Cisco, Renesas, Quectel, and AWS throughout a 25-year career in very diverse hardware and software roles. He has his own lab at home (including a 3D printer), which some university colleagues praised as "more complete and prepared than the ones we had when we studied."

His other passions are his family, his dog, and playing and coaching underwater hockey.

It amazes me how much things have changed since I first connected a lightbulb, a switch, and a battery, or wrote "10 PRINT CHR$(147)".

Coding on a computer was for "hackers" then; these days kids learn programming at primary school.

Making a 1-layer PCB required drawing skills and handling perchloric acid; today you can get a 4-layer design assembled and delivered to your door.

And learning? There are so many resources at "clicking" distance.

Love it, let's do more!

Table of Contents

3

IoT End Devices, the Neuron Cells of an IoT System   35

4

Wireless Connectivity, the Nervous Pathway to Delivering IoT Data   57

5

The Cloud, IoT's "Superpower Brain"   89

Part 2: Utilizing AI in IoT Development

6

Applying ChatGPT in the IoT Innovation Journey   107

9

Using AI Tools to Draw Application Flow Diagrams   157

Part 3: Practicing an End-to-End Project

10

Setting Up the Development Environment for Your First Project   171

11

Programming Your First Code on ESP32   197

16

Creating a Data Visualization Dashboard on ThingsBoard   301

Preface

Are you overflowing with innovative ideas, yet finding it difficult to navigate the intricacies of software coding? Utilize **Artificial Intelligence (AI)** to expedite your **Internet of Things (IoT)** development journey!

AI is a transformative force that is reshaping our lives, societies, and industries. This book guides beginners on how to use AI's coding abilities to construct their first end-to-end IoT prototype. It covers everything from drawing an application flow diagram, crafting the hardware prototype, producing embedded C++ example code, establishing Wi-Fi connectivity, and accessing **Amazon Web Services (AWS)**, to creating a real-time dashboard on ThingsBoard Cloud.

The book ensures a smooth learning curve, starting from the IoT fundamentals, architecture, key elements, recommendations, and best practice examples, to a thorough step-by-step hands-on project illustration. A distinguishing feature of this book is its exploration of recent AI advancements and their transformative impact on the IoT world. It emphasizes ChatGPT prompt skills specifically tailored for IoT projects and presents a detailed framework for crafting effective ChatGPT prompts. This empowers you to harness this powerful tool in your IoT endeavors, overcoming barriers related to inadequate software coding skills or experience.

You will be introduced to the PlatformIO IDE on Visual Studio Code, one of the most popular embedded software development environments. Additionally, you'll learn about the cutting-edge RISC-V architecture MCU – ESP32, Arduino-compatible sensors, and integration methods for the AWS cloud and ThingsBoard dashboard.

As part of the learning approach, I provide the functional codes generated by ChatGPT and prompting instruction examples in the GitHub repo for this book.

By the end of this book, you will be equipped to build your first successful IoT prototype, effectively bridging the gap between your innovative ideas and functional creations.

Who this book is for

This book is designed for beginners who are eager to explore the world of IoT technology but face obstacles due to limited experience in embedded software coding, particularly in C++. The primary audience includes middle- to high-school and undergraduate students, hobbyists interested in smart home applications, hardware enthusiasts, DIY creators, startup entrepreneurs, educators, and professionals from non-technical backgrounds. Often, their innovative potential is hindered by the complexity of software coding. Fortunately, AI can serve as an intelligent assistant, offering example code to accelerate the development of IoT prototypes.

This book assumes that readers have a basic understanding of electronic physics, knowledge of internet and IP connectivity, a rudimentary grasp of software coding structure, and familiarity with basic cloud concepts.

What this book covers

- *Chapter 1, IoT Essentials, All You Should Know*, offers an introduction to the core concepts and components that define IoT technologies.

- *Chapter 2, IoT Network, the Neural System of Things*, gives you a detailed look at the infrastructure and architecture of IoT systems.

- *Chapter 3, IoT End Devices, the Neuron Cells of an IoT System*, helps you understand the role and functionality of the devices at the edge of IoT networks.

- *Chapter 4, Wireless Connectivity, the Nervous Pathway to Delivering IoT Data*, explores the wireless technologies that convey data within IoT systems.

- *Chapter 5, The Cloud, IoT's "Superpower Brain"*, discusses the capabilities and advantages of cloud computing in enhancing and scaling IoT applications.

- *Chapter 6, Applying ChatGPT in the IoT Innovation Journey*, explores how AI can assist in conceptualizing, planning, and executing IoT solutions, emphasizing the transformative impact of AI on the innovation process.

- *Chapter 7, Recommendations to Start Your First IoT Project*, provides you with practical advice on initiating your first IoT projects.

- *Chapter 8, 10 Beginner-Friendly IoT Projects with ChatGPT Prompts*, details 10 projects using ChatGPT to various extents, showcasing how AI can streamline the development process, from sensor integration to data handling.

- *Chapter 9, Using AI Tools to Draw Application Flow Diagrams*, teaches you about the use of AI-driven tools to create IoT application flow diagrams.

- *Chapter 10, Setting Up the Development Environment for Your First Project*, provides a detailed guide on how to establish a development environment using Visual Studio Code and the PlatformIO IDE.

- *Chapter 11, Programming Your First Code on ESP32*, walks through ChatGPT-assisted C++ programming on an ESP32 microcontroller to read sensor data locally.

- *Chapter 12, Establishing Wi-Fi Connectivity*, explains the steps to configure the ESP32 to connect to a Wi-Fi network.

- *Chapter 13, Connecting the ESP32 to AWS IoT Core*, illustrates the configuration of AWS IoT Core settings, certificate management, and how to establish a secure MQTT/TLS connection from the ESP32.

- *Chapter 14, Publishing Sensor Data to AWS IoT Core*, guides you through the process of setting up sensors, collecting data, and publishing it to the cloud.

- *Chapter 15, Processing, Storing, and Querying Sensor Data on AWS Cloud*, has you practicing tasks to process data via Python functions on AWS Lambda and store and query data using AWS IoT Analytics.

- *Chapter 16, Creating a Data Visualization Dashboard on ThingsBoard*, explains how to integrate the AWS cloud with ThingsBoard, set up real-time data feeds, and build your own dashboards.

To get the most out of this book

You will need to have an understanding of the basics of software development IDE tools, open source hardware, and the cloud.

Software/hardware covered in the book	Operating system requirements
PlatformIO IDE extension on VS Code	macOS and Windows
ESP32 MCU	
Arduino-compatible sensors	
AWS services	MacOS, Windows, Linux(Any)
ThingsBoard Cloud	MacOS, Windows, Linux(Any)

Download the example code files

You can download the example codes and ChatGPT prompting files for this book from GitHub at `https://github.com/PacktPublishing/Accelerating-IoT-Development-with-ChatGPT`. If there's an update to the code, it will be updated in the GitHub repository.

Conventions used

There are a number of text conventions used throughout this book.

`Code in text`: Indicates code words in text, database table names, folder names, filenames, file extensions, pathnames, dummy URLs, user input, and Twitter handles. Here is an example: "Utilizing `build_flags` in this context offers several benefits."

A block of code is set as follows:

```
 1. [env:esp32-c3-devkitc-02]
 2. platform = espressif32
 3. board = esp32-c3-devkitc-02
 4. framework = arduino
 5. monitor_filters = esp32_exception_decoder, colorize
 6. monitor_speed = 115200
 7. build_src_filter = +<../../src/>  +<./>
 8. board_build.flash_mode = dio
 9. build_flags =
10.     -DARDUINO_USB_MODE=1
11.     -DARDUINO_USB_CDC_ON_BOOT=1
12.     -w
13. lib_deps =
14.     adafruit/DHT sensor library@^1.4.6
15.     adafruit/Adafruit Unified Sensor@^1.1.14
```

Any command-line input or output is written as follows:

```
pio platform install espressif32
```

Bold: Indicates a new term, an important word, or words that you see onscreen. For instance, words in menus or dialog boxes appear in **bold**. Here is an example: "The ESP32 will reboot and you can locate the **Serial Monitor** button."

> **Tips or important notes**
> Appear like this.

Get in touch

Feedback from our readers is always welcome.

General feedback: If you have questions about any aspect of this book, email us at customercare@ packtpub.com and mention the book title in the subject of your message.

Errata: Although we have taken every care to ensure the accuracy of our content, mistakes do happen. If you have found a mistake in this book, we would be grateful if you would report this to us. Please visit www.packtpub.com/support/errata and fill in the form.

Piracy: If you come across any illegal copies of our works in any form on the internet, we would be grateful if you would provide us with the location address or website name. Please contact us at copyright@packt.com with a link to the material.

If you are interested in becoming an author: If there is a topic that you have expertise in and you are interested in either writing or contributing to a book, please visit authors.packtpub.com.

Share your thoughts

Once you've read *Accelerating IoT Development with ChatGPT*, we'd love to hear your thoughts! Scan the QR code below to go straight to the Amazon review page for this book and share your feedback.

https://packt.link/r/183546162X

Your review is important to us and the tech community and will help us make sure we're delivering excellent quality content.

Download a free PDF copy of this book

Thanks for purchasing this book!

Do you like to read on the go but are unable to carry your print books everywhere?

Is your eBook purchase not compatible with the device of your choice?

Don't worry, now with every Packt book you get a DRM-free PDF version of that book at no cost.

Read anywhere, any place, on any device. Search, copy, and paste code from your favorite technical books directly into your application.

The perks don't stop there, you can get exclusive access to discounts, newsletters, and great free content in your inbox daily

Follow these simple steps to get the benefits:

1. Scan the QR code or visit the link below

https://packt.link/free-ebook/978-1-83546-162-4

2. Submit your proof of purchase
3. That's it! We'll send your free PDF and other benefits to your email directly

Part 1: Understanding IoT Fundamentals

In this section, you will explore in depth the fundamental components of IoT systems, using an enlightening analogy with the human neural system to illustrate the functions and interconnected nature of these technologies. You will gain a solid understanding of the essential concepts that underpin the IoT ecosystem, learning about the critical roles of IoT networks and end devices, and the various options for wireless connectivity that act like the nerves in a body, delivering data seamlessly across distances. Moreover, you will study how the cloud functions as the brain of IoT systems, providing substantial computing power and vast storage capabilities to process and manage data efficiently. This section sets the stage for a deeper appreciation of how IoT operates within and interacts with various applications.

This part contains the following chapters:

- *Chapter 1, IoT Essentials, All You Should Know*
- *Chapter 2, IoT Network, the Neural System of Things*
- *Chapter 3, IoT End Devices, the Neuron Cells of an IoT System*
- *Chapter 4, Wireless Connectivity, the Nervous Pathway to Delivering IoT Data*
- *Chapter 5, The Cloud, IoT's "Superpower Brain"*

1

IoT Essentials, All You Should Know

In this chapter, you will go through the concepts and definitions of the **Internet of Things (IoT)** as defined by the industry, as well as its remarkable evolution milestones since its inception. By understanding the historical context and outlook of IoT, you will gain valuable insights into its significance and potential applications in the market.

Furthermore, you will zoom into the wide-ranging market that IoT addresses and various industries and sectors where its impact is most prominent. From healthcare to transportation, from smart homes to industrial automation, IoT has revolutionized the way we live, work, and interact with our environment. You will explore these real-world applications and the transformative power of IoT in shaping our society.

By the end of this chapter, you will have acquired a solid foundation and a well-rounded understanding of essential concepts and knowledge required to embark on your journey into the intricate world of IoT. You will be fully prepared to explore the details and intricacies of IoT as we dive deeper into the subject matter in subsequent chapters of this book.

In the engineering projects in *Chapter 11*, with the assistance of ChatGPT, you will learn to program C++ code on an ESP32 **microcontroller unit (MCU)**, send diverse sensors' data through your home Wi-Wi network, and store, analyze, and visualize live and dynamic sensor data on the **Amazon Web Services (AWS)** cloud. In this chapter, we will acquaint ourselves with these terms so as to build the right foundation.

This chapter covers the following topics:

- The evolving definition of IoT
- Addressable markets
- How IoT impacts us

The evolving definition of IoT

IoT has never stopped evolving its definition since its debut in the 1990s.

According to Gartner (`https://www.gartner.com/en/information-technology/glossary/internet-of-things`), the concept of IoT is defined as follows:

> *"The Internet of Things (IoT) is the network of physical objects that contain embedded technology to communicate and sense or interact with their internal states or the external environment."*

However, as of today, this definition may seem a little cliché and out of date. Initially, the notion of IoT was specifically designed to support traditional **machine-to-machine** (**M2M**) communication within manufacturing plants, where it was constrained by wired copper twists, Ethernet cables, and power supply cords. It was typically deployed in a fixed scenario with no mobility, scalability, elasticity, power consumption, and cost efficiency.

Since 1999, when IoT was officially named by Kevin Ashton, MIT's Executive Director of Auto-ID Labs, it has been making significant strides in various areas. This is thanks to breakthroughs in industrial innovations such as silicon chipsets, wireless technologies, and cloud services. IoT no longer operates solely on a local network but goes beyond massive geographical areas on a vast scale. IoT is now expanding its reach beyond homes, residential and commercial areas, campuses, and cities, and is even reaching deserts, barren zones, oceans, and aerospace.

Looking back, the evolution of IoT has experienced several remarkable leaps since its inception. These leaps include improvements in deployment coverage, power consumption, cost structure, and architecture, as described next.

Deployed from fixed to mobility

The year 2000 was a pivotal moment for the IoT explosion. It marked a significant milestone in the advancement of IoT applications. This was made possible by the industrial innovation of low-cost 802.11 Wi-Fi networks, which were rapidly and widely adopted in both homes and enterprises. The introduction of the 802.11 Wi-Fi network brought about a revolution for IoT devices. They were no longer solely reliant on cables (although a power cable might still be necessary), but could now benefit from the convenience of mobility. This groundbreaking development enabled IoT devices to be seamlessly connected through wireless connectivity in residential, commercial, and business spots, whether they were stationary or on the move. This was a monumental shift in the IoT landscape, paving the way for even greater possibilities and innovations in the future.

802.11 Wi-Fi is not the only option dominating the home and enterprise spaces; there are other new players that provide local mobility as well, including the 802.15.1 family of Bluetooth and **Bluetooth Low Energy** (**BLE**) and the 802.15.4 family comprising ZigBee, Thread, and Matter, and Sub-GHz Z-Wave.

Mobility from a local area to a wide area

The year 2005 was a significant catalyst for boosting IoT with the introduction of **service provider**'s (**SP**) 3G cellular networks, including **Code-Division Multiple Access** (**CDMA**) and the **Universal Mobile Telecommunications System** (**UMTS**). These networks offered faster and wider area mobility connectivity, enabling a greater variety of devices to connect to the internet at will. This breakthrough has opened up numerous opportunities for IoT applications in various areas, including beyond residential and campus, spanning cities and rural areas. Some examples of these applications include fleet management, smart agriculture, and asset tracking.

Heading into the year 2010, the introduction of 4G/**Long-Term Evolution** (**LTE**) technology accelerated the growth of IoT by providing higher data rates. In 2020, the era of 5G brought even more rapid progress. As a result, IoT has penetrated almost every population corner of the world, transforming the way we live and work.

Presence from city to barren fields

The wide area expansion of the IoT footprint is not solely tied to the coverage of traditional 4G/LTE and 5G cellular networks provided by SPs. In less populated areas such as barren fields, forests, mountains, offshore, and even oceans, where SPs are hesitant to offer coverage due to unwise investment, industries are adopting new wireless technologies to provide IoT coverage. Examples of such technologies include **low-power wide area networks** (**LPWANs**), such **as Long-Term Evolution Category M** (**LTE-M**), **Narrow Band Internet of Things** (**NB-IOT**), **long-range wide-area network from LoRa Alliance** (**LoRaWAN**), **Sigfox**, and **Low Earth Orbit** (**LEO**) networks, such as SpaceX Starlink and Amazon Kuiper, which do not rely on traditional SPs.

LTE-M and NB-IoT are both derived from SPs' cellular networks. Unlike traditional 4G/LTE and 5G technologies, which strive to provide high data rates to mobile phones, LTE-M and NB-IoT prioritize wider coverage and superior lower power consumption for battery-powered IoT devices with reduced receiving sensitivity, bandwidth, and data rates, which are distinct from conventional 4G/LTE and 5G networks.

LoRaWAN is an industrial standard supported by the LoRa Alliance. It is designed to provide long-range coverage and low power consumption for various IoT applications. By operating on **unlicensed spectrum** (such as the band allocated for **Industrial, Scientific, and Medical** (**ISM**)), LoRaWAN offers accessibility to a wide range of users who are interested in utilizing it for their IoT projects. This standard ensures that devices can communicate over long distances while consuming minimal power, making it an ideal choice for many IoT scenarios.

SpaceX Starlink and Amazon Kuiper LEO satellite constellations will be instrumental in facilitating the deployment of IoT applications in those areas. With their extensive coverage and high-speed connectivity, these satellite networks will empower various IoT devices and enable seamless data transmission, enhancing communication and connectivity for users in remote and underserved regions.

Furthermore, the integration of SpaceX Starlink and Amazon Kuiper LEO with IoT technologies will unlock new possibilities for industries such as agriculture, transportation, and environmental monitoring, revolutionizing the way we collect, analyze, and utilize data for improved decision-making and efficiency.

Throughput from Mb/s to Gb/s

When IoT was initially designed for M2M communication in manufacturing plants, it was commonly accomplished using RS232 or RS485 wire interfaces. The twisted wire limited the data rate from a few hundred kilobits per second up to 10 Mb/s. However, with the emergence of advanced technologies such as Wi-Fi 5, Wi-Fi 6, 4G/LTE, and 5G, the maximum data rate has significantly increased to a range of over 100 Mb/s to 1 Gb/s. This tremendous enhancement in data transfer capabilities has opened up a multitude of new possibilities for IoT applications.

One of the major benefits of these advancements is the ability to support high-definition video surveillance. With the increased data rate, IoT devices can now transmit high-quality video feeds, allowing for better monitoring and security. This is particularly useful in areas such as public spaces, where surveillance is crucial for ensuring safety.

Another exciting possibility enabled by these advancements is the immersive experiences they can provide. With the higher data rate, IoT devices can now support technologies such as **augmented reality (AR)** and **virtual reality (VR)**. This means that users can have more interactive and engaging experiences, whether it's through AR apps that overlay digital information onto the real world or VR simulations that transport them to virtual environments.

Battery life from days to years

Chasing a high data rate for IoT applications is not always necessary. While some IoT applications do require high data rates, it is important to consider that there are also many instances where external power input is not available. In these cases, the IoT applications must be powered by batteries that need to last for several years. As a trade-off, these applications can accept very low data rates ranging from a few bytes per second to a few hundred kilobits per second, and have higher latency of up to a few seconds.

Recognizing this need, in addition to continue exploring high data rate technology, industries also shifted their focus toward developing technologies offering low data rates and low power consumption. This has led to the emergence of various wireless technologies that cater to these requirements. LoRaWAN, which was mentioned earlier, is one such example. Other notable technologies include NB-IOT and LTE-M, which are offered by SPs through their cellular networks. Additionally, there is **Bluetooth Low Energy Long Range (BLE-LR)**, and 802.11 ah Wi-Fi Halow.

Networking from point-to-point and point-to-multipoint to mesh and star topologies

Traditional M2M communication, such as RS232 (point-to-point) and RS485 (point-to-multipoint) has limitations when it comes to the scale of nodes and communication distance. In order to overcome these limitations and expand IoT connectivity beyond manufacturing plants to residential, community, city, and rural areas, industries have developed mesh topologies. These mesh topologies connect IoT devices as child nodes and bridge to each other to deliver data to parent nodes, enabling a wider coverage area. Examples of mesh topologies include BLE and ZigBee meshes for smart homes and Wi-SUN meshes for utility smart metering.

By implementing mesh topologies, the scale and coverage of IoT networks have been significantly extended. However, this expansion also brings forth new challenges, such as increased networking complexity, less efficient usage of radio resources, and higher device power consumption. Despite these challenges, the benefits of mesh topologies in terms of connectivity and coverage make them a viable solution for some specific use cases, as mentioned previously.

To address these challenges and meet the growing demands of industries, various networking approaches have been explored. One widely used option is the star topology, which offers networking simplicity, such as LoRaWAN, NB-IOT, LTE-M, and LEO satellite networks.

In a star topology, the IoT device can easily connect to an access device with just one radio hop. This access device can take the form of a LoRaWAN gateway, the base station (that is, eNodeB) in a cellular network, or even a LEO satellite. This type of networking is known for its **Plug and Play** (**PnP**) functionality, making it easy to set up and use. Furthermore, it allows IoT devices to conserve power by entering deep sleep mode when they are not actively executing tasks, thereby reducing overall power consumption compared to a mesh topology where every device needs to be always on to bridge the network.

Intelligence at edge node

In the era of M2M communication, IoT devices, known as slave nodes, collect and transmit untapped protocol data to their upstream nodes, such as the RS232 or RS485 master node. The main function of these IoT devices is to forward raw data without locally possessing intelligence. However, industries have recognized the benefits of developing local computing capabilities for IoT devices. This includes using containers on top of the device OS, Google TensorFlow, and AWS IoT Greengrass.

Edge intelligence provides significant advantages to IoT by processing untapped data locally using predefined algorithms instead of sending all raw data upstream indiscriminately. This approach leads to savings in backhaul bandwidth consumption, reduced application decision latency, and offloading work from the cloud.

While the benefits of edge intelligence are significant, it is important to note that it may not be suitable for every IoT application. This is due to increased requirements for IoT device hardware design, such as the need for a powerful processor, larger memory, and high-power consumption.

AI and ML

The evolution of intelligence does not stop at IoT devices alone. Industries are continuously pushing **artificial intelligence** (**AI**) and **machine learning** (**ML**) capabilities into the IoT cloud. By leveraging AI and ML algorithms, cloud platforms such as AWS and Microsoft Azure empower businesses to analyze and make sense of vast amounts of IoT data. These platforms provide advanced tools and services that enable organizations to develop intelligent applications, automate processes, and make data-driven decisions.

One key advantage of integrating AI and ML into the cloud is the scalability and flexibility it offers. With the ability to scale resources on demand, businesses can easily handle the increasing volume, velocity, and variety of IoT data. This ensures that organizations can process and analyze data in real time, enabling them to respond quickly to changing conditions and make informed decisions.

Furthermore, AI and ML algorithms can uncover patterns, trends, and correlations in IoT data that might otherwise go unnoticed. This allows businesses to gain valuable insights into customer behavior, operational efficiency, and predictive maintenance, among other things. By leveraging these insights, organizations can optimize processes, improve customer experiences, and drive innovation.

In addition to processing and analyzing IoT data, cloud platforms also provide the infrastructure and tools needed to train and deploy AI models. This enables businesses to build intelligent applications that can learn and adapt over time, improving their performance and accuracy. With services such as Amazon SageMaker and **Azure Machine Learning** (**AML**), organizations can easily develop, train, and deploy AI models without the need for extensive computational resources or expertise.

Moreover, cloud-powered AI and ML bring the benefits of accessibility and affordability. By leveraging the cloud, businesses of all sizes can access cutting-edge AI and ML capabilities without the need for significant upfront investments in infrastructure or specialized talent. This democratization of AI allows organizations to level the playing field and compete in an increasingly data-driven world.

With its exponential growth and rapid expansion from its initial purpose, IoT has ambitiously embraced and integrated various innovative technologies within a relatively short span of time. Its primary objective is to cater to an extensive market, extending beyond just homes and campuses to industries, cities, and even rural areas. By doing so, IoT seeks to revolutionize and transform the way we live, work, and interact with technology on a global scale, which proactively extending its addressable markets from residential, neighborhood, campus, city and rural areas. Let's learn more about it next.

Addressable markets

When exploring the IoT addressable market, there are several approaches available to make segmentation possible, such as differentiating from technology specifications, application designs, business models, or applicable verticals. However, this chapter will focus on an easy-to-understand market segmentation approach based on the application's premises. This approach can be especially beneficial for beginners to precisely identify their target market for innovation. By understanding where applications are being deployed, beginners can gain valuable insights into the specific needs and challenges of various locations.

Residential

A smart home is a convenient and accessible environment for beginners to practice their first IoT innovations. It provides a low-hanging fruit scenario for newbie developers to experiment with their innovation.

One major advantage of a smart home is its easy access to existing home Wi-Fi networks, eliminating the need for additional network infrastructure. Additionally, the power supply is not a concern as IoT devices can be powered by DC power from wall sockets. Even for IoT devices that require batteries, replacing them is effortless when needed.

There are already several major vendors that offer a central hub solution for hosting smart home applications, including Google Home, Amazon Alexa, and Apple HomeKit. Beginners can follow these vendors' specifications to ensure compatibility and seamless integration of their IoT innovations into mobile applications and cloud platforms. This is particularly important for the recent smart home interworking protocols of **If This Then That** (**IFTTT**) and Matter.

The following is a list of the top 10 popular smart home applications:

- Indoor temperature and humidity monitoring
- Smoke/flame detectors
- Security cameras
- Smart lighting/bulbs
- Smart thermostats
- Smart locks
- Smart doorbells
- Smart wall power socket
- Smart garage door opener
- Smart trash bin

Without a doubt, the primary focus for IoT deployment is at home due to its significant market scale and easy **go-to-market** (**GTM**) approach. In comparison, IoT applications designed for commercial and business markets present higher requirements for product quality, management, reliability, and security, along with the potential for higher profit margins.

Security concerns pertaining to the use of IoT applications within the home environment primarily revolve around potential breaches of privacy. These concerns arise from the fact that these IoT devices, which are increasingly becoming pervasive in our daily lives, are often connected to the internet and, therefore, carry the inherent risk of sensitive personal information being exposed to unauthorized individuals or entities.

Commercial and business

IoT applications in commercial and business locations can be easily extended from smart home applications. For instance, a smoke detector can be used in both homes and business buildings. However, accessing enterprise Wi-Fi networks poses challenges due to the security mechanisms implemented for enterprise-grade access. While accessing a home Wi-Fi network requires simply using the Wi-Fi **service set identifier** (**SSID**) and password, accessing an enterprise Wi-Fi network requires support for **Wi-Fi Protected Access 2** (**WPA2**), 802.1X **Extensible Authentication Protocol** (**EAP**), and **Remote Authentication Dial-In User Service** (**RADIUS**) authentication.

When it comes to installing and managing IoT devices within a business location, there are two usual challenges. Firstly, finding an external power supply above the floor is not always easy. You cannot simply lay power cords inside the building as you please. Secondly, if you are deploying hundreds of devices, you will need to create a management solution to effectively monitor and manage them.

The following is a list of the top 10 popular IoT applications within commercial and business locations:

- Indoor temperature and humidity monitoring
- Smoke/flame detectors
- Security cameras
- Smart lighting/bulbs
- Smart **Heating, Ventilation, and Air Conditioning** (**HVAC**) management
- Smart parking
- Asset tracking and inventory management
- Water leakage detection
- Building intrusion detection
- Occupancy detection

The IoT applications designed for commercial and business markets mainly focus on indoor deployment, while at its adjacent segment, penetrating the community and campus market has to overcome challenges from the deployment of IoT applications in open spaces, the outdoor case, which raises the bar to the product's robustness and reliability.

Securing IoT applications in commercial and business environments involves regular device updates, strong authentication, secure boot processes, and **hardware security modules** (**HSMs**) to protect devices. Network security is enhanced through encryption, network segmentation, firewalls, and **intrusion detection systems** (**IDSs**). Data security measures include encrypting data at rest and in transit, strict access controls, and data minimization. Cloud security involves secure APIs, compliance with regulations, and regular security audits. Physical security measures include access control and tamper detection.

Neighborhood and campus

Most IoT applications in neighborhood and campus settings are primarily focused on outdoor use cases. The process of building IoT applications in neighborhood and campus environments poses greater challenges compared to commercial and business premises. These challenges include the need for outdoor ruggedized design, reliable wireless connectivity, and power supply availability.

When IoT devices come to outdoor hardware designs, they adhere to the **Ingress Protection standard (IP grade)**. This standard ensures that the hardware is capable of withstanding various environmental factors and conditions, such as dust, water, and extreme temperatures. By following the IP grade, manufacturers can ensure that their outdoor hardware is durable, reliable, and suitable for outdoor deployments. Whether it is for smart city infrastructure, rural area applications, or neighborhood and campus settings, outdoor hardware needs to be designed with the utmost consideration for protection and resilience. The details will be explained in the *Device types* section of *Chapter 3*. Wireless connectivity in the neighborhood and campus poses a challenge. Ideally, a public 802.11 Wi-Fi network should be able to meet the requirements. However, IoT devices must support WPA2, 802.1X EAP, and RADIUS authentication, as with the security access mechanism used in enterprise networks. In case a public 802.11 Wi-Fi network is not available, alternative networks should be explored.

One option is to access the 4G/LTE or 5G network provided by local SPs. To do this, your IoT device module needs to have a compatible cellular module. You will also need to purchase a SIM card and data plan from the SPs. Another option is to access the LoRaWAN network hosted by local public LoRaWAN operators. The worst-case scenario is that you build your own private LoRaWAN network, although this option can be complex and costly.

Another major challenge in neighborhood and campus settings is the availability of an external power supply for IoT devices. It can be difficult to find an accessible power source or to lay down power cables to the installation location. To address this issue, it is recommended to design IoT devices with low power consumption, incorporating solar panels or rechargeable batteries.

The following is a list of the top 10 popular IoT applications within neighborhood and campus locations:

- Smart pathway lighting
- Smoke/flame detectors
- Security cameras
- Smart parking
- Smart water irrigation
- Smart health care
- Asset/pet tracking
- Smart waste management

- Wireless emergency button
- Facility security

The neighborhood and campus markets need to support IoT applications either by fixed installations or local mobility within a limited area. In most cases, enterprise Wi-Fi and private LoRaWAN networks are the preferred cost-effective options. 4G/LTE and 5G can also be considered, but cost is always the top concern. For the city market, in addition to cost, wide area mobility, seamless coverage, and business scalability also matter.

In terms of security considerations, neighborhood and campus spaces are significantly more vulnerable to potential attacks. This increased vulnerability is due to the openness of these spaces, which allows users to bring their own IoT devices. For example, a user might bring a device such as a pet tracking tag. These devices, while useful and convenient, can inadvertently introduce security risks into the network, thereby increasing the potential for unauthorized access and attacks. Therefore, it's crucial to have robust security measures in place to mitigate these risks and protect the network.

Cities

When IoT applications are deployed in cities, they can be seen as an expanded version of community and campus environments, but on a larger scale and with more challenges. The main challenges are ensuring wide area mobility, seamless coverage, and business scalability. To address these challenges, one option is to utilize a 4G/LTE or 5G network provided by local SPs or access the public LoRaWAN network from local LoRaWAN operators.

The following is a list of the top 10 popular IoT applications within cities:

- Smart street lighting
- Noise detection
- Security cameras
- Smart parking
- Smart water irrigation
- Connected car
- Asset/pet tracking
- Smart waste management
- Air quality monitoring
- Facility security

The city market is the most complex and challenging segment, especially due to its geographic space. The rural market is also a challenging one, as it often lacks cellular network coverage despite having IoT applications deployed over large areas.

Security considerations for IoT deployment in cities are also a multifaceted challenge. As cities become increasingly interconnected and reliant on IoT technologies for their functioning, the importance of ensuring the security of these systems escalates significantly. Foremost among security concerns is the complexity arising from the sheer number of interconnected devices. Cities deploying IoT technologies may be dealing with millions of devices, each representing a potential point of vulnerability. These devices, ranging from traffic sensors to smart meters, all need to be secured to prevent malicious actors from exploiting them, which is a daunting task due to their sheer number and diversity. Data security is another major concern. The vast amount of data generated by IoT devices needs to be securely transmitted, stored, and processed. This data often includes sensitive personal information, making its security a key priority. Any breach in data security could have severe implications, ranging from privacy violations to significant operational disruptions.

Rural areas

IoT applications deployed in rural areas are most relevant to agriculture, the natural environment, and wildlife tourism. In most cases, there is a lack of cellular network coverage. The options include accessing the public LoRaWAN network from local LoRaWAN operators, building your own private LoRaWAN network, or accessing LEO satellite networks. However, both hardware costs (**CapEx**) and data plans (**OpEx**) are always a concern.

The following is a list of the top 10 popular IoT applications within cities:

- Soil moisture monitoring
- Wildfire detection
- Air quality monitoring
- River water pollution monitoring
- Flood monitoring
- Landslide alerting
- Animal tracking
- Fleet management
- Security cameras
- Facility security

The rural area market is witnessing the flourishing of IoT applications related to agriculture production, environment protection, and natural disaster-proofing. However, wireless connectivity remains a top concern that needs to be addressed. As with the industrial market, there are numerous options available for resolving this issue.

Security considerations for rural areas are often less complex than those for cities. The primary reason behind this is that rural areas are likely to have fewer types of IoT applications. For example, in a city, IoT technology might be used for everything from traffic management systems to smart home applications and environmental monitoring. In contrast, in a rural setting, the applications might be more limited, perhaps focusing mainly on agriculture or weather monitoring. This limited scope can result in fewer security concerns, simply because there are fewer potential points of vulnerability. However, it's still crucial to have robust security measures in place, regardless of the complexity or scope of the IoT deployment.

Industries

IoT is also an indispensable component of Industry 4.0. It empowers a multitude of industrial processes such as smart manufacturing, predictive maintenance, and **supply-chain optimization** (SCO). Through the collection and analysis of real-time data, IoT enhances automation, improves energy management, ensures quality control, and enables customized production.

However, the integration of IoT in industrial environments is not without challenges. These include ensuring robust security measures to protect against cyberattacks, achieving interoperability among a diverse range of devices, managing vast volumes of data, and scaling systems efficiently. Balancing the costs of implementation, addressing privacy concerns, obtaining the necessary technical expertise, and maintaining reliability in demanding industrial environments are also significant hurdles to overcome.

Despite these challenges, overcoming them is of paramount importance for industries to fully leverage the potential of IoT. The increased efficiency, innovation, and competitiveness that IoT offers can revolutionize industrial processes and significantly enhance productivity and profitability.

In the industrial market, IoT applications are often unique, leading to a wide variety of wireless connectivity options across different verticals. Depending on the specific needs and requirements of the industry, various wireless connectivity options can be utilized. These may include Wi-Fi 6, 6E, and 7, private 4G/LTE such as **Citizens Broadband Radio Service** (CBRS), 5G **Ultra-Reliable Low Latency Communications** (URLLC), **WiSUN** mesh, **ISA100**, and **WirelessHART** for more specialized applications.

The following is a list of the top 10 popular IoT applications within industries:

- Smart metering
- **Supervisory Control and Data Acquisition (SCADA)**
- Automotive mobile robotics
- Security cameras

- Pipeline leakage monitoring

- Preventive maintenance

- Fleet management

- Asset tracking

- **Hazardous location (Hazloc)** monitoring

- Facility security and people safety

The **industrial IoT (IIOT)** market is unique compared to other IoT markets because each vertical, such as utilities, mining, oil and gas, manufacturing, and transportation, has its own specific requirements. This requires special expertise from IoT application developers in these industries. This situation raises the bar for market penetration.

We are seeing the widespread use of IoT applications in many markets, such as homes, cities, and industries. This expansion is not only changing our immediate surroundings but also influencing various aspects of our lives. In the next section, we will discuss the real effects of this technological revolution. We will look at how IoT is changing convenience and efficiency in our homes, promoting sustainable and smart solutions in our cities, and transforming processes and safety measures in industries. Each of these areas shows the diverse impact of IoT, highlighting its significant role in the era of digital transformation.

How IoT impacts us

Following the stride of IoT expanding into new spaces, it consistently impacts us from every corner, every dimension, and every time. Some of them you are just aware of, with tangible benefits that catch your attention remarkably; some of them you are ignoring because they have already been part of your daily life for many years.

In summary, there are four key improvements that have been brought about by the expansion of IoT into various spaces. These improvements have had a significant impact on our lives, making them more comfortable, safe, efficient, and environmentally friendly.

Living comfort and safety improvement

In our daily lives, IoT has brought about numerous benefits, greatly enhancing our comfort and safety within various areas of our living spaces.

Personal care

Personal care is one of the most popular IoT applications among individuals, typically using a smartwatch to measure heart rate, blood oxygen level, walking steps, and location. With the widespread adoption of BLE and the streamlined Matter protocol, increasingly portable IoT devices are being connected to mobile devices and forwarding real-time data to the cloud platform.

In the past, your grandpa may have measured his blood pressure daily and then written the value down in his notebook. He would then bring the recorded daily data to his family doctor during monthly visits. While IoT technology makes it more convenient and enables easy data delivery to family doctors, there are new innovative blood pressure monitors in the market that have BLE connectivity to mobile phone apps and then to cloud platforms. The collected daily data is recorded and aggregated in mobile phone apps and forwarded to the cloud. Family doctors can have real-time visibility into people's blood pressure management situation and can even be promptly alerted by abnormal events.

Today, thanks to low-power-consumption MCUs, sensors, and wireless technologies, battery-powered IoT applications are increasingly emerging in the field of personal care, especially for the elderly, children, and individuals with disabilities. While smart wearables remain the most popular option in the market, we are also witnessing other innovations; for example, a BLE- and LoRaWAN-equipped wireless emergency button on necklaces for the elderly can trigger alerts anywhere to associated caregivers in case of emergencies; a location tracking tag on children's backpacks that transmits GPS information to parents' mobile apps through ubiquitous 4G/LTE networks; a blind cane equipped with ultrasound sensors to detect distance to obstacles in front of blind people and sound an alarm if they are too close.

Sweet home

Home is a cozy place to relax and unwind, and it has also become a shining stage where IoT is actively dancing. Thanks to home Wi-Fi connectivity, people now have the power to remotely control a wide range of smart home appliances with just a few taps on their smartphones. This technological advancement has truly transformed the way we interact with our homes.

Imagine having a smart thermostat that not only measures the room temperature in real time but also automatically adjusts the home HVAC system to your desired level of comfort. Gone are the days of manually adjusting the thermostat; now, you can simply use your smartphone to ensure that your home is always at the perfect temperature. Forgot to turn off the HVAC when you left home? No worries! You can either remotely turn it off through mobile apps, or your smart thermostat system can automatically turn it off when it detects that nobody is staying in the room. But that is not all. With a smart power wall outlet, you can schedule your home appliances to turn on or off at specific times, allowing you to save electricity and reduce your carbon footprint. Gone are the days of accidentally leaving appliances on; now, you can have peace of mind knowing that your home is energy-efficient and environmentally friendly. And let us not forget about smart lighting bulbs. These innovative lighting solutions can automatically switch on or off based on ambient brightness. Say goodbye to fumbling around in the dark or wasting energy by leaving lights on during the day.

Furthermore, IoT has greatly improved safety in residential spaces. By leveraging diverse IoT technologies, home security systems have become more intelligent, offering a wider range of features to ensure the security and peace of mind of homeowners. These systems now include wireless surveillance cameras, motion sensors, and smart windows and door locks, which work together to provide a comprehensive security solution. Through internet connectivity and mobile phone apps, users can easily access live video feeds from their surveillance cameras, receive instant notifications about any suspicious activity detected by the motion sensors, and even remotely lock or unlock their doors. More smart security-related applications are coming from smart home industries. For example, traditionally, home smoke detectors would only sound alarms locally within the house. However, with IoT integration, smoke detectors can now send alerts to your mobile phone apps through the internet, allowing homeowners to receive notifications about potential fire hazards even when they are away from home. This added layer of connectivity and convenience ensures that homeowners can take immediate action to prevent any potential disasters.

Operational efficiency improvement

Premises such as businesses, commercials, merchandise stores, school campuses, and cities serve as battlegrounds for IoT to fight against operational inefficiency.

Energy saving

Energy consumption at these locations is the largest factor contributing to operational costs. This includes costs related to electricity, water, and gas. With the advent of IoT technology, there are innovative solutions available to optimize electricity consumption specifically for lighting and HVAC systems. By implementing IoT technology, business owners can improve energy efficiency and reduce costs. With IoT, ceiling lighting and HVAC systems can be intelligently controlled to work according to the actual demands of the building. This means that the lights and air conditioning will only be turned on when needed, rather than being constantly on.

A real-life example that demonstrates the application of smart energy technology is in meeting rooms. In these spaces, IoT sensors are utilized to detect human activity. When these sensors detect that there is no activity in the meeting room, the system is automatically triggered to dim the lighting and air conditioning, reducing their intensity or turning them off completely.

Predictive maintenance

Commercial buildings are equipped with a wide array of machines and equipment that are critical for their proper functioning and the safety and comfort of the people inside. These machines include water pumps, gas valves, ventilation systems, air conditioning compressors, and many more. Any malfunction or failure in these systems can have profound consequences. However, with the help of IoT sensors, it is now possible to monitor and track the performance of these machines in real time. These sensors can detect any anomalies or deviations from normal operating conditions and predict potential malfunctions or failures before they occur. This allows for proactive and preventive maintenance measures to be taken, ensuring that any issues are addressed before they escalate and cause disruptions. By implementing IoT sensors, commercial buildings can greatly improve their operational efficiency, reduce downtime, and enhance the overall safety and comfort of their occupants.

A real-life example is the use of vibration monitoring applications on air conditioning compressors. In this case, an IoT sensor is attached to the compressors to record vibration range data. The sensor then sends the real-time data to the cloud. The system administrator can access online visibility to monitor the operational status of the compressor and will be alerted if the data analysis predicts the need for maintenance.

Another case in commercial buildings is the detection of water leakage. This is a frequent problem that can lead to severe damage if not addressed promptly. To address this issue, water leakage detection sensors can be strategically installed at the corners of the room where water pipelines are more prone to leaks. These sensors continuously monitor for any drops of water leakage and immediately alert the building administrators in case of any. This proactive approach ensures that any potential water leakage is identified and dealt with swiftly, minimizing the risk of extensive damage and providing a safe and secure environment for everyone using the meeting rooms.

Environmental protection improvement

Open field spaces such as natural areas, forests, national parks, mountains, rivers, offshore areas, and oceans are remarkable occasions for adopting IoT applications.

Disaster warning

Disaster monitoring is an extremely important aspect of environmental protection. The ability to effectively monitor and respond to disasters is crucial for saving lives and minimizing damage. In areas where there is often no cellular coverage, innovative IoT applications utilizing alternative wireless backhauls, such as LoRaWAN and LEO satellites, have emerged as key players in disaster monitoring efforts.

A real case is a volcano activity monitoring application through vibration sensors with LoRaWAN wireless backhaul. These vibration sensors are strategically installed in the volcano area to continuously monitor the vibration data of rocks in real time. This valuable data is then transmitted to a cloud platform where scientists and researchers can access and analyze it. By utilizing this advanced technology, experts are able to closely observe and study volcanic activity, enabling them to make more accurate predictions and take necessary precautions to ensure the safety of nearby communities and infrastructure.

Another example of how IoT technology is being used is in the detection of wildfires in California. Flame detection sensors have been installed in areas prone to wildfires. If a fire starts, these sensors will be triggered and send an alarm to the cloud, allowing the government to take immediate action. This is an important development as it helps in the early detection and prevention of wildfires, ultimately protecting lives and property.

Pollution monitoring

Pollution control is one of the significant benefits brought by IoT to environmental protection. In addition to disaster monitoring, IoT technology has facilitated the development and implementation of various smart sensors for pollution detection. These sensors can detect a wide range of pollutants, including but not limited to carbon dioxide (CO_2), ammonia (NH_3), nitrogen oxides (NO_x), alcohol, benzene, smoke, particulate matter (PM2.5), acetone, thinner, formaldehyde, and water quality. By utilizing these smart sensors, environmental agencies and organizations can effectively monitor and mitigate pollution levels, leading to a healthier and safer environment for all.

A real-life example to consider is the implementation of air quality sensors on the factory chimney. These highly advanced sensors are designed to accurately measure the quality of exhaust emissions. By doing so, they provide valuable data that can be seamlessly transmitted to the government's central monitoring system. This allows for continuous monitoring and analysis of the factory's environmental impact, aiding in the enforcement of regulations and the implementation of necessary measures to ensure improved air quality.

A similar case is river water quality monitoring sensors. By installing these sensors at the riverside, particularly near factories or residential areas, the government can effectively monitor the quality of river water. This monitoring system plays a crucial role in ensuring the health and safety of both the environment and the people living nearby. Additionally, these sensors provide valuable data that can be used for further analysis and decision-making regarding water management and pollution control measures.

Industrial productivity improvement

Industries such as manufacturing, utilities, oil and gas, mining, transportation, and agriculture are always significantly benefited by IoT technologies.

Cost savings

One of the major advantages of implementing IoT is the significant cost savings, particularly in terms of labor costs. This cost reduction can be attributed to the automation and optimization of various processes and tasks made possible by IoT technology. By leveraging IoT devices and systems, businesses can streamline their operations, eliminate manual and time-consuming tasks, and increase efficiency. This not only leads to cost savings but also allows for the reallocation of resources to other areas of the business, fostering growth and innovation.

A real case is the **Advanced Metering Infrastructure (AMI)** in the utilities industry. There are millions of electricity meters within a state in the US, and it is super costly to assign technicians to read such scale of meters on every street and every month. Nowadays, IoT has introduced various wireless technologies, such as 802.15.4 mesh, for collecting data from those meters. In other countries, utility companies are now using LoRaWAN, NB-IoT, and LTE-M for the same purpose.

Another case is smart water irrigation in the agriculture business. To conserve water during irrigation, farmers currently install soil moisture sensors in the ground. These sensors measure the water saturation level and report it to the customer service center. If the water saturation meets the requirements of the plants, the irrigation system will be stopped. Additionally, the customer service center will determine the daily water irrigation volume based on the local climate forecast. If rain is predicted for the following day, they will adjust the water irrigation volume accordingly.

Scale expansion

Wireless IoT technology is revolutionizing the industrial scale by eliminating the constraints of cables, providing greater flexibility and convenience for industrial customers. With its ability to connect various devices and collect real-time data, wireless IoT technology enables businesses to optimize their operations, improve efficiency, and make informed decisions. Additionally, this technology opens new possibilities for automation, remote monitoring, predictive maintenance, and enhanced safety measures in industrial settings. The widespread adoption of wireless IoT technology is transforming industries and driving innovation across sectors, leading to increased productivity, cost savings, and improved overall performance.

One example of a real case is the implementation of **Autonomous Mobility Robotics (AMR)** in various settings such as manufacturing plants, warehouses, and logistic distribution centers. IoT has made it more convenient to provide Wi-Fi, 4G/LTE, and 5G connectivity to AMR units at any location where customers wish to establish a new site.

Summary

In this chapter, we embarked on a journey through the fundamental concepts of IoT, gained insights into its vast and ever-expanding market, and explored how IoT seamlessly integrates into various markets, from our personal spaces in homes to the broader realms of cities and industries. The chapter also shed light on the profound impact of IoT, illustrating how it is not just reshaping our interactions with technology but also revolutionizing our day-to-day efficiency, sustainability, and overall quality of life. By understanding the basics and recognizing the major influence of IoT, we set the stage for delving deeper into each of its architecture's composed elements and got ready to practice the 10 engineering projects listed in *Chapter 11*, including hardware illustration, ChatGPT prompt templates, and example codes.

Further reading

There are many professional books talking about IoT from different perspectives, the following book is highly recommended which were written my former talent colleagues.

IoT Fundamentals: Networking Technologies, Protocols, and Use Cases for the Internet of Things, David Hanes, Gonzalo Salgueiro, Patrick Grossetete, Robert Barton, Jerome Henry, Cisco Press.

It's a good idea to acquaint yourself with key terms associated with IoT before moving forward. Following are some of them:

Term	Explanation
IoT	**Internet of Things**, a network of interconnected devices that can collect and exchange data using embedded sensors, software, and other technologies.

Term	Explanation
Gateway	An aggregate device that stays between the end device and the internet, providing secured wired or wireless connectivity and data transport protocol support.
End device	An individual device at the end of the IoT network to generate data or execute commands, usually comprising an MCU, sensors, actuators, and wireless backhaul.
MCU	Microcontroller unit, a compact process in an embedded system.
Sensor	A device that detects and responds to some type of input from the physical environment, such as temperature, light, pressure, and so on.
Actuator	A device responsible for moving or controlling a mechanism or system, often receiving an operational indication from an upstream control system.
Node	A node is typically a physical device or a data point in a network that can send, receive, or route data.
Edge computing	A method of optimizing cloud computing systems by performing data processing at the edge of the network, near the source of the data.
Cloud computing	A model that enables easy and convenient access to a shared pool of configurable computing resources (such as networks, servers, storage, applications, analytics, and intelligence services) through the internet.
Wireless backhaul	The use of wireless communication systems to transport data from remote network edges to a central network, or the core of the network.
4G/LTE	The fourth generation of the cellular network standard, offering faster data transfer rates than 3G networks.
5G	The fifth generation of the cellular network standard, offering higher speeds and reduced latency compared to 4G/LTE.
NB-IOT	**Narrowband IoT**, a low-power **wide area network** (**WAN**) radio technology standard from 4G/LTE that focuses on wide coverage, low cost, long battery life, and high connection density.
LTE-M	**Long-Term Evolution for Machine**, a type of low-power WAN radio technology from 4G/LTE standard tailored for IoT, allowing for longer battery life and better indoor coverage.
Wi-Fi	A wireless technology that provides wireless network connections, following the *IEEE 802.11* family of standards, to various devices such as computers, smartphones, and smart home devices without physical wired connections.
Wi-Fi 6	802.11ax Wi-Fi.
Wi-Fi 6E	An extension of Wi-Fi 6, operating in the 6 GHz frequency band, providing faster speeds and lower latencies.
Wi-Fi 7	802.11be Wi-Fi.

Term	Explanation
Wi-Fi HaLow	802.11ah Wi-Fi, a Wi-Fi specification developed for IoT; its main distinction lies in its use of sub-1 GHz license-exempt bands.
LoRa	A spread spectrum modulation technique derived from **chirp spread spectrum (CSS)** technology, designed for long-range, low-power communication between IoT devices.
LoRaWAN	A low-power, wide-area networking **Media Access Control (MAC)** layer protocol based on the LoRa modulation, designed for wirelessly connecting battery-operated "things" to the internet in regional, national, or global networks, and is commonly used in IoT applications such as smart cities and IIoT.
BLE	**Bluetooth Low Energy**, a wireless personal area network technology designed and marketed for novel applications in healthcare, fitness, beacons, security, and home entertainment.
BLE-LR	**Bluetooth Low Energy Long Range**, an extension of BLE, optimized for long-range connectivity.
ZigBee	A low-power, wireless mesh networking technology, based on the *IEEE 802.15.4* standard, commonly used for connecting simple, smart devices such as home automation systems and IoT sensors.
Matter	An open source, royalty-free connectivity standard that enables reliable, secure communication among a wide range of smart devices.
Thread	A low-power, wireless mesh networking protocol designed for IoT devices to communicate on a local network.
MQTT	**Message Queuing Telemetry Transport**, a lightweight, **publish-subscribe (pub/sub)** network protocol, ideal for the minimal-bandwidth, high-latency communication required in various IoT applications, such as remote monitoring.
Raspberry Pi	A small and all-in-one computer that you can use to learn programming and for practical projects, widely used in IoT for its versatility.
Arduino	An open source electronics platform, known for its accessible software and versatile hardware boards, extensively utilized in education, prototyping, and DIY (Do It Yourself) projects, enabling enthusiasts and professionals alike to develop interactive gadgets that can sense and manipulate the physical environment.
RISC-V	An open standard **instruction set architecture (ISA)** based on established **reduced instruction set computer (RISC)** principles, often used in microprocessors for IoT devices due to its flexibility and adaptability.

2

IoT Network, the Neural System of Things

The IoT network typically consists of IoT end devices, wireless connectivity (in most cases), and a backend cloud platform. It operates like the neural system of the human body, which consists of neuron cells, nervous pathways, and the brain. Neuron cells serve as biological sensors, detecting stimuli from the skin or muscles and generating biological electrical signals, similar to IoT end devices sensoring the object status change. These biological electrical signals travel through nervous pathways, which are comparable to delivering sensor data through wired/wireless connectivity in an IoT network.

Eventually, these biological electrical signals reach the brain, the central command center, as with the cloud platform in IoT. Just as the brain interprets, processes, and responds to sensory information, cloud services collect, store, analyze, and make decisions based on the received data from IoT end devices.

In this chapter, you will start from the multiple options of considering proper IoT network options for your target use case. This will help you make informed technology decisions for different deployment locations, whether at home, on campus, in a building, city, or even in a rural area.

Please note that IoT networks on industrial premises are unique and require specialized knowledge. This book will not cover this specific topic. Still, it is important to acknowledge that implementing and designing IoT systems in industrial settings has its own challenges and considerations. These may include network scalability, service reliability, data security, and integration with existing infrastructure.

This chapter covers the following topics:

- IoT networks at home
- IoT networks on campuses and in buildings
- IoT networks in cities
- IoT networks in rural areas

IoT networks at home

The best practice for setting up an IoT network at home typically involves using **home Wi-Fi**, **BLE**, and **Thread**. When designing IoT applications for home use, such as monitoring temperature and humidity, smoke detectors, security cameras, and pathway lighting, it is important to prioritize simplicity, affordability, and ease of use for customers. Consequently, the architecture is specifically designed to meet the needs of residents, providing a friendly and enjoyable user experience.

Home Wi-Fi

Wi-Fi, short for **Wireless Fidelity**, is a low-cost wireless communication technology that enables devices to connect to the internet and communicate with each other without physical cables. It complies with *IEEE 802.11* standards and utilizes unlicensed radio frequency bands, typically 2.4 GHz, 5 GHz, and 6 GHz (Wi-Fi 6E). The home Wi-Fi network follows a star topology, which offers much simplicity and easier installation. In most cases, a home Wi-Fi router can serve up to 30 Wi-Fi client devices simultaneously.

Home Wi-Fi allows devices such as smartphones, tablets, laptops, smart TVs, and smart home devices to connect to the internet. Wi-Fi offers higher throughput ranging from hundreds of Mbps to Gbps, enabling users to access online services, browse the web, stream media, and use video surveillance and home security applications without the need for physical cables. By connecting to the internet, Wi-Fi enables control and monitoring of various smart devices in your home from anywhere in the world, using your smartphone, tablet, or laptop.

However, Wi-Fi technology is not power consumption friendly compared to BLE, ZigBee, and LoRaWAN. Wi-Fi has strong transmission capability and wall penetration ability. It has the most powerful performance, but the power consumption of Wi-Fi is so high that it is not suitable for battery-powered devices and is more suitable for power plug-in devices. Some may argue that the presence of 802.11ah, also known as Wi-Fi HaLow, is intended to provide low-power-consumption connectivity. However, it is specifically targeted for use cases that are not within residential areas.

We will discuss more details about Wi-Fi technology in *Chapter 4*.

BLE

BLE is a wireless technology that provides a low data rate, making it ideal for devices that require power efficiency. Its primary purpose is to establish connections with battery-powered devices, including but not limited to wearable devices and healthcare devices.

In the realm of smart home technology, devices such as Amazon Alexa Echo, Google Home/Nest, Apple HomePod, and Apple TV serve as BLE hubs. These hubs play a vital role in the IoT network within a home, allowing for seamless connectivity and control over a wide range of smart devices in a home environment. These smart devices can include lighting systems, thermostats, security cameras, and much more.

By utilizing BLE technology, these smart home hubs ensure reliable and efficient communication between devices, enhancing the overall smart home experience. Users can easily manage and control their smart devices through voice commands or mobile applications, creating a convenient and personalized living space.

We will discuss more details about BLE technology in *Chapter 4*.

Thread

Thread, based on the *IEEE 802.15.4* standard, complies with mesh networking technology that connects smart home devices, from motion sensors to door locks, and offers a better, more efficient way for them to work together. In a Thread network, multiple devices can act as board routers or mesh extenders, expanding the network's reach and reliability. Unlike traditional star topology networks, where signals are only sent back to the original sender, mesh networks repeat and forward signals to other devices within range, creating a signal-hopping characteristic. This allows for greater range and reliability, as devices can relay signals to each other, extending the network's coverage. Additionally, mesh networks automatically expand as more devices are added, resulting in a stronger and more reliable network that can support a growing number of smart home devices.

Thread natively adopts IPv6 technology, which offers numerous advantages for IoT applications, such as smart home, smart campus, and smart buildings applications. It benefits from well-established networking security and seamlessly integrates into the existing network infrastructure. Unlike legacy automation systems that operate in separate networks for each application, Thread enables multiple applications, even those based on different standards and protocols, to share the same network simultaneously. Additionally, each individual device can utilize the most suitable physical network type. Thread's low-power mesh network is particularly suitable for power consumption-sensitive devices that need to cover large areas within a location.

It is worth mentioning that there is currently a significant development in the application layer of Thread. The **Matter protocol**, formerly known as **Project CHIP (Connected Home over IP)**, is being adopted to enhance the functionality and interoperability of Thread-based devices. Matter aims to create a unified standard for smart home devices, ensuring seamless integration and ease of use for consumers.

Matter adopted in the application layer of Thread signifies a major milestone in the evolution of smart home technology. With the adoption of the Matter protocol, manufacturers and developers can now leverage a standardized framework to build Thread-based devices that are compatible with a wide range of smart home ecosystems. This not only simplifies the development process but also promotes innovation and collaboration within the industry.

In addition, the adoption of Matter brings numerous benefits to consumers. Establishing a unified standard for smart home devices eliminates the issue of compatibility between varied brands and platforms. Consumers can now confidently purchase Thread-based devices, knowing that they will seamlessly integrate with their existing smart home setup. This not only enhances convenience but also encourages the widespread adoption of smart home technology.

The following figure compares the stack difference to Wi-Fi, Thread, Bluetooth Classic and Low-Energy.

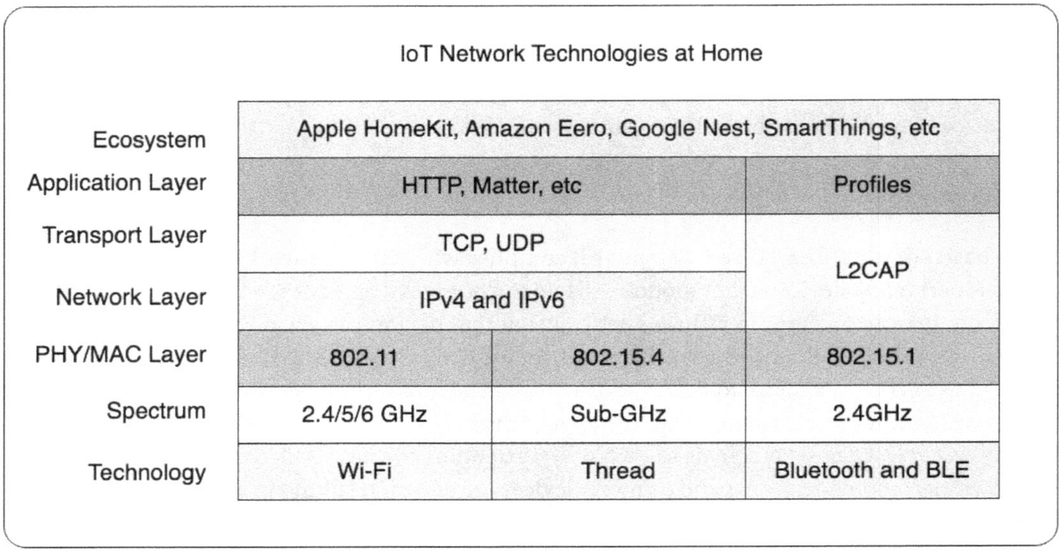

	IoT Network Technologies at Home		
Ecosystem	Apple HomeKit, Amazon Eero, Google Nest, SmartThings, etc		
Application Layer	HTTP, Matter, etc		Profiles
Transport Layer	TCP, UDP		L2CAP
Network Layer	IPv4 and IPv6		
PHY/MAC Layer	802.11	802.15.4	802.15.1
Spectrum	2.4/5/6 GHz	Sub-GHz	2.4GHz
Technology	Wi-Fi	Thread	Bluetooth and BLE

Figure 2.1 – IoT network stack at home

In addition to the IoT network at home, another popular case is on campuses and in buildings. Let's learn more about it.

IoT networks on campuses and in buildings

The best practice for deploying IoT networks on campuses and in buildings typically includes **enterprise Wi-Fi**, a **Thread mesh**, a **private LoRaWAN network**, and specific industrial networks such as Wi-SUN mesh for electricity metering and street lighting. Although BLE and ZigBee mesh may have a limited presence in these locations, they are not widely utilized. This section will discuss popular IoT coverage options provided by enterprise Wi-Fi, Thread for buildings, and private LoRaWAN networks on campuses and in buildings.

Enterprise Wi-Fi

IoT applications on campuses and in buildings typically utilize an enterprise-grade architecture. This architecture considers factors such as indoor/outdoor connectivity, security, scalability, management, and cost. The most common wireless network used for IoT applications in these settings is enterprise Wi-Fi.

Compared to home Wi-Fi access, it is more than critical to consider and implement a robust security mechanism that allows end devices to access the enterprise Wi-Fi access point. Once the end devices have access to the enterprise Wi-Fi network, data is transmitted through the enterprise intranet, which is a private network. From there, it is forwarded either to the application platform at the enterprise data center or to the platform in the public cloud.

Enterprise Wi-Fi at campus and building locations usually offers two topology options: star and mesh.

A star topology is a network configuration where all devices are connected to a central hub, as known as a Wi-Fi access point. This hub serves as the main access point for routing data traffic. The star topology is known for its straightforward design and easy management. When a device needs to communicate with the internet, it sends the data to the central hub, which then forwards it to the intended destination. This straightforward setup makes troubleshooting and maintenance easier. A star topology is well suited for smaller enterprises and business spots that prioritize simplicity and ease of management, such as small offices, restaurants, and merchant franchises where a single switch connects all Wi-Fi client devices.

A Wi-Fi mesh is a network where multiple Wi-Fi access points function as a unified whole. It is well suited for large campuses and building locations. Coordinated Wi-Fi access points are strategically placed throughout a campus, building, or larger area to provide access to the same wireless network. Depending on the manufacturer and technology chosen, one of these access points may serve as the main device, often referred to as a controller, or they can all be considered equal.

Most Wi-Fi mesh networks share the following characteristics:

- **Single SSID**: This feature allows for a single network name and password. Wi-Fi client devices can automatically roam among the mesh network without the need for additional SSID provisioning.

- **Client steering**: Wi-Fi client devices that connect to the network are automatically associated with the best radio-quality access points around them.

- **Band steering**: The network also determines the optimal frequency band to provide the best performance for each client at any given time.

- **Self-healing**: If a Wi-Fi access point becomes temporarily or permanently unavailable, the network will automatically reroute traffic to avoid the issue.

A mesh topology is exceptionally reliable and redundant in network connectivity. In the event of a link or access point's failure, data can be rerouted through alternative paths, ensuring uninterrupted connectivity. However, this strength comes with the trade-off of increased complexity and cabling requirements.

WPA2 for Enterprise, also known as **WPA2-ENT**, was launched in 2004 and is still widely considered the standard for wireless network security. It offers encryption over the air and a higher level of security. When combined with the effective authentication protocol called 802.1X, users can be authorized and authenticated for secure network access. Deploying WPA2-ENT requires a RADIUS server, which handles the task of authenticating network user access. The authentication process is based on the 802.1X policy and uses various systems labeled EAP. By authenticating each device before connection, a personal, encrypted tunnel is effectively established between the device and the network.

Thread mesh

The concept of Thread for home will be explained in the last section. It is highlighted that for the expansion of the Thread network to smart buildings, along with the utilization of a Thread border router and battery-operated devices, it is typically necessary to incorporate the implementation of **Thread Mesh Extender (TME) devices**. This addition ensures a scalable deployment of the network, enabling seamless connectivity and enhanced performance throughout the building infrastructure.

As a mesh network, Thread enables devices to not only receive data but also actively participate in routing data to other devices, known as TME devices. This unique capability of Thread creates a highly robust and efficient network infrastructure. TMEs, as the name suggests, are always-on Thread devices that extend the range and link quality of the Thread network by acting as routers for messages between other devices, such as battery-powered sensors. By allowing devices to act as relays, the network expands its coverage area without the need for additional repeaters or extenders. As a result, Thread provides a stable and reliable network with extensive reach, ensuring seamless connectivity for all devices within the network.

Private LoRaWAN network

In addition to Wi-Fi, there is another wireless technology called LoRaWAN that can be utilized for IoT applications on campuses and in buildings. LoRaWAN is specifically designed for low data rates, long-range connectivity, and low power consumption. It is particularly suitable for applications that require long-range coverage and low power consumption. With LoRaWAN, IoT devices can be deployed in large areas such as campuses and buildings without the need for frequent battery replacements or complex infrastructure setups.

LoRaWAN end devices communicate with single or multiple LoRaWAN gateways over long distances using an unlicensed radio spectrum known as the **industrial, scientific, and medical (ISM)** band. Unlike Wi-Fi, which operates on the 2.4 GHz and 5 GHz ISM bands, LoRaWAN usually operates on the Sub-GHz band. In Europe, it uses the 868 MHz frequency, while in the US, it uses 915 MHz. This Sub-GHz band offers better penetration and propagation capabilities than Wi-Fi, making it especially suitable for providing coverage deep inside buildings.

The architecture of LoRaWAN is unique. Behind the LoRaWAN gateway, the data arrives at the LoRaWAN network server first. The network server is responsible for device management, payload encryption, and the application API.

The following is a standard LoRaWAN network diagram used on campuses and in buildings, including both open-field outdoor and indoor coverage:

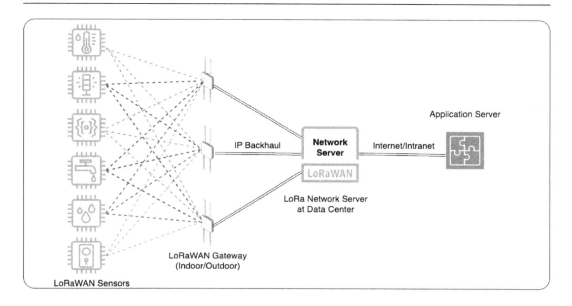

Figure 2.2 – LoRaWAN private network

The LoRaWAN network implemented in these locations follows a loosely coupled star topology. Unlike enterprise Wi-Fi, a LoRaWAN endpoint does not establish a predefined association with a dedicated LoRaWAN gateway. Instead, the data transmitted from a LoRaWAN endpoint, also known as an uplink message, can be received by several LoRaWAN gateways in the neighboring area. However, the returning data, known as **downlink messages**, will be delivered from the best-signal-quality LoRaWAN gateway to the endpoint. The LoRaWAN gateway is typically installed on the wall or ceiling indoors and on a lighting pole outdoors.

A private LoRaWAN network is a model in which the entire infrastructure is deployed and hosted by the IoT application providers. In this model, the IoT application providers take the responsibility of installing the LoRaWAN gateway themselves, ensuring seamless and efficient network connectivity for their IoT devices. With a private LoRaWAN network, the IoT application providers have complete control over the network operations and can tailor them to their specific needs and requirements. This allows for greater flexibility and customization in managing and monitoring connected devices, ensuring optimal performance and security.

No one-size-fits-all approach

When it comes to IoT applications, it is important to understand that Wi-Fi, Thread, and LoRaWAN are not interchangeable options. Each of these wireless technologies has its own strengths and considerations.

Wi-Fi is a smart choice for IoT applications that require high throughput, such as surveillance cameras. It offers fast and reliable connectivity, but it does come with a requirement for an external power supply.

On the other hand, Thread and LoRaWAN are specifically designed for use cases that prioritize low power consumption. While they may have lower throughput compared to Wi-Fi, devices equipped with Thread and LoRaWAN radio can typically operate on batteries for years. This makes them an ideal choice for devices such as smoke detectors, water leakage detectors, and security sensors, as well as temperature and humidity sensors.

The IoT network on campuses and in buildings can be easily built and hosted by enterprise customers themselves. Extending IoT networks to cover the entire city demands seamless mobility and higher availability to serve multiple customers and diverse applications, which will increase complexity.

IoT networks in cities

Applications of IoT in a city can be seen as a complex system on a large scale. This system has different parts and technologies that work together to create an efficient and seamless system. One of the main challenges in implementing IoT in a city is making sure there is good wireless network coverage. This is important for IoT devices to work well and exchange data. The best practice to set up an IoT network in a city is usually by using a cellular network (such as 4G/LTE and 5G) and a public LoRaWAN network from public IoT SPs. These technologies offer wide coverage and can reach deep into buildings, which makes them great for IoT solutions in cities.

Cellular network

4G/LTE and 5G networks use cellular coverage from SPs to provide reliable connectivity for IoT devices from LTE-M and NB-IoT with low power consumption and low data rate (LPWAN), and superior high data rate technologies such as **LTE Advanced (LTE-A)** and 5G **Enhanced Mobile Broadband (eMBB)**. These networks are well established and widely available, making them a popular choice for diverse IoT deployments. IoT devices on this network need a cellular modem and SIM/eSIM chip. These devices communicate with eNodeB, which is the base station of the cellular network. Data is then sent from eNodeB to the SP's packet core network and finally reaches the data platform, which can be in a data center or the cloud.

The cellular access network typically has a star topology. SPs install eNodeB stations on cellular towers and city building roofs. These strategically placed eNodeB stations are crucial for communication with end devices, which have cellular modems and SIM cards or eSIM chipsets.

The following is the standard architecture of a 4G/LTE network:

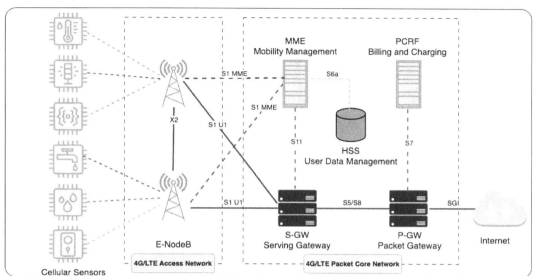

Figure 2.3 – 4G/LTE network

Public LoRaWAN network

LoRaWAN uses unlicensed spectrum, making it a radio resource-sharing network. Unlike cellular networks, which assign dedicated air resources to each end device, LoRaWAN aims to support use cases on a best-effort basis. As a result, it may not be suitable for all IoT applications, especially those that are mission-critical. The topology of the LoRaWAN network in a city is the same as the one deployed on campuses and in buildings, which is a loosely coupled star topology as well. In this topology, there is a central hub called the LoRaWAN gateway that acts as a bridge between the end devices and the network server. The LoRaWAN gateway can be installed in various locations within the city. It can be mounted on the roof of a city building, providing wide coverage and a clear line of sight for communication. Another option is to install it on an electricity power pole, strategically placing it for optimal signal transmission. Lastly, the LoRaWAN gateway can also be mounted on a street light pole, taking advantage of existing infrastructure to expand the network reach.

When thinking about how well a network can handle lots of users, stay available, and be reliable, a public LoRaWAN network usually uses a cloud-based LoRaWAN network server. This cloud-based server can easily grow and adapt to handle more users and higher demands.

The following is a cloud-based LoRaWAN network diagram:

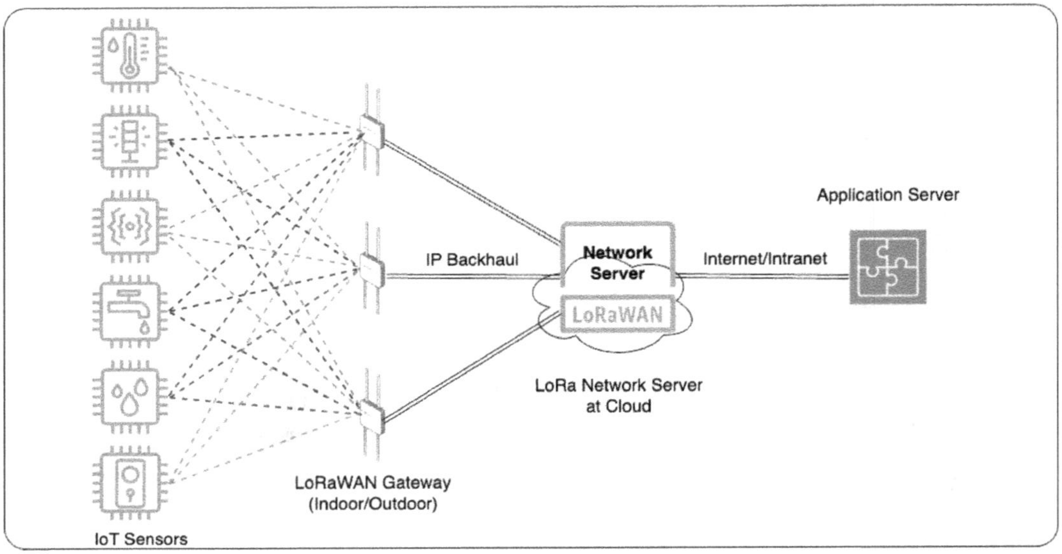

Figure 2.4 – Cloud-based public LoRaWAN network

AWS IoT Core (https://aws.amazon.com/iot-core/) for LoRaWAN is a fully managed LoRaWAN network server. It allows you to connect and manage LoRaWAN devices using LoRaWAN connectivity with the AWS cloud. With AWS IoT Core for LoRaWAN, customers can establish their LoRaWAN network by connecting many LoRaWAN devices and gateways to the AWS cloud, without having to develop or operate a network server for LoRaWAN themselves. This removes the burden of managing a network server and enables easy and fast connection and security for LoRaWAN device fleets at scale.

By using the long-range and wide in-building coverage of LoRaWAN, AWS IoT Core for LoRaWAN helps customers speed up IoT application development. This is achieved by utilizing data generated from connected LoRaWAN devices through AWS services.

IoT networks in cities can leverage the 4G/LTE cellular network or other public-hosted wireless infrastructure such as LoRaWAN. In rural areas, any public coverage rarely reaches there; that is a significant challenge, but there still is a chance.

IoT networks in rural areas

Those spaces, such as agriculture, oil fields, mining sites, forests, national parks, mountains, coastlines, and offshore areas, usually do not have cellular network coverage. IoT applications targeted for these areas must overcome the challenges of wireless connectivity. The best practice for accessing an IoT network in addition to a cellular network in a rural area is a **private LoRaWAN** network and a LEO network.

Private LoRaWAN network

In addition to deploying LoRaWAN networks on campuses and in buildings, establishing a private LoRaWAN network in rural areas is also a reliable and cost-effective option. With its long-range capabilities and low power consumption, LoRaWAN is ideal for various IoT applications. In such cases, IoT application providers have the flexibility to create their own private LoRaWAN coverage at the location where they offer services. This model allows them to have complete control over the network and ensure reliable connectivity for their IoT devices.

By establishing a private LoRaWAN network, IoT application providers can overcome the limitations of relying on external SPs and ensure seamless connectivity for their devices. They can customize the network according to their specific requirements, ensuring maximum coverage and performance.

LEO network

Another good option for getting wireless connectivity in rural areas for IoT applications is to use a wireless broadband network provided by LEO satellites, such as Starlink. With this approach, you just need to set up a Starlink terminal at the location. This terminal acts as a gateway and gives you a reliable Wi-Fi hotspot to connect your IoT devices.

Using LEO satellites helps overcome the limitations of traditional cellular networks or local infrastructure for IoT deployments. The Starlink terminal plays an important role by connecting your IoT devices to the satellite network, allowing them to communicate seamlessly over long distances. This solution is especially useful in remote or hard-to-reach areas where cellular coverage is limited or not available.

Summary

In this chapter, we explored the wide-ranging world of IoT networks. We saw how they are used in different settings: making our homes more convenient, improving efficiency on campuses and in buildings, driving innovation in cities, and enhancing rural areas.

As we move on to the next chapter, we shift our focus from the entire network to individual elements: the end device. In this chapter, we will look closely at the different types of devices that are the building blocks of an IoT network. We will study the hardware architecture of these devices, with a special focus on MCUs, which are like the brains of these devices. We will also discuss peripherals and interfaces that allow the devices to interact with their surroundings, as well as sensors and actuators that make the environment responsive and interactive.

3

IoT End Devices, the Neuron Cells of an IoT System

Just as neurons are the basic building blocks of the human neural system, the fundamental component in an IoT network is the IoT end device. These devices, like the sensing function of neural cells, are designed to monitor the surroundings, report changes, and perform actions.

In this chapter, you will learn about the crucial role of IoT end devices in IoT applications. These devices detect, capture, measure, and report the status or data changes of target objects to backend cloud platforms. They can also execute operations on these objects based on commands received from the backend.

To give you a comprehensive understanding of the functions and design of IoT end devices, we will discuss them from various perspectives. These include the common categories of IoT end device types, as well as their hardware architecture, MCUs, peripherals, input and output interfaces, and the sensors and actuators commonly used in IoT applications.

By the end of this chapter, you will have a well-rounded understanding of the fundamental knowledge required for IoT end-device design. You will also be equipped with the necessary skills to create a hardware prototype that fits your specific use case requirements, enabling you to effectively implement your first IoT solutions.

This chapter covers the following topics:

- Device types
- Hardware architecture
- MCUs
- Peripherals and interfaces
- Sensors and actuators

Device types

A variety of IoT devices are available in the market. Categorizing them can be challenging due to their diverse functions and purposes. These devices are designed to support different application needs and requirements. This section aims to classify the devices based on popular distinctions, aiding beginners in developing a better understanding. Recognizing these distinctions is crucial when crafting your own IoT innovations, as these are pervasive and should not be overlooked.

Installing indoors versus outdoors

An IoT end device is installed either indoors or outdoors. Indoor installations include ones in homes, buildings, public facilities, and industrial factories. Outdoor installations encompass ones done in open fields on campuses, city open spaces, outskirts, rural farm fields, forests, mountains, and barren areas. The deployment location of IoT end devices should be chosen carefully based on their application. Factors such as chipset selection, mechanical enclosure design, operational temperature and humidity ranges, and device mount options need to be considered.

When installing IoT devices in residential spaces such as homes, apartments, hotels, and classrooms, which are called consumer-grade devices, you do not need a waterproof or strict dustproof design. A lightweight, durable plastic enclosure is usually enough. The devices can work in temperatures ranging from 0°C to 40°C (32°F to 104°F) and in humidity levels ranging from 5% to 95%. There are also other indoor environments such as offices or warehouses, where the devices are known as commercial-grade products. In these places, the devices need to work in temperatures ranging from -20°C to 70°C (-4°F to 158°F). You can install IoT devices indoors in three ways: on the wall, on the ceiling, or by sticking them up.

When IoT devices are designed to be used outdoors, such as in open fields, downtown city streets, or rural areas, they are usually called **outdoor-grade devices**. The hardware design requirements for these devices are stricter compared to indoor devices. Outdoor devices must have a much wider operational temperature range compared to indoor devices, typically ranging from -40°C to +85°C (-40°F to 185°F).

To withstand tough outdoor conditions, outdoor devices must consider additional factors such as the **Ingress Protection Grade (IP grade)** for their enclosures.

Usually the IP grade complied by a product is described by two digits as the given condition. We will learn more about this in the following table.**First digit**	**Mechanical Protection**	**Second digit**	**Water ingress protection**
0	No protection	0	No protection

1	Protected against solid objects over 50mm such as hands	1	Protected against vertically falling drops of water such as condensation
2	Protected against solid objects over 12mm such as fingers	2	Protected against direct sprays of water up to 15° from the vertical direction.
3	Protected against solid objects over 2.5mm such as tools and wires	3	Protected against direct sprays of water up to 60° from the vertical direction.
4	Protected against solid objects over 1mm such as wires, nails, and so on	4	Protected against water splashed from all directions, limited ingress permitted
5	Protected against limited dust ingress, not harmful deposits	5	Protected against low-pressure jets of water from all directions, limited ingress permitted
6	Completely protected against dust	6	Protected against strong jets of water such as on the deck of the ship, limited ingress permitted
N/a	N/a	7	Protected against the effects of temporary immersion between 15cm and 1m; the duration of the test is 30 minutes
N/a	N/a	8	Protected against extended periods of immersion under pressure

Table 3.1 – IP grade definitions

In certain open-field locations, such as high-altitude areas and coastal areas, UV resistance and anti-salt fog enclosures are also needed in addition to the IP grade.

These outdoor devices also offer diverse ways to be mounted, such as mounting on a lighting pole, attaching to a cable, mounting on a roof, or mounting on a tower. When installing these devices outside, it is important to remember that the mount options may need to have anti-gust capabilities.

Powering via external supply versus batteries

An IoT device can operate on either an external power supply or a battery, making it suitable for both indoor and outdoor usage. Devices such as surveillance cameras, wireless speakers, and street lightbulbs typically request high power consumption and are often powered externally. This power can be supplied through an AC/DC power adapter or converter.

While these devices are more complex and costly, they offer advantages such as continuous service, high reliability, and low latency. They also support **edge computing**, which demands more power due to its robust computing capability and quick memory access.

As power consumption is not a significant concern for devices with an external power supply, they can utilize high-speed wired connectivity (Ethernet, Coax Cable, or Fibers) or high-throughput wireless connectivity (Wi-Fi, LTE CAT-4 or above, or even 5G).

Battery-operated devices are rapidly expanding and offer significant benefits for their intended use. They are portable and can be moved around freely, making them ideal for IoT applications that can tolerate low data rates, service interruptions, and delays.

These devices typically feature MCUs and wireless modules that are designed for minimal power consumption. To conserve energy, they often switch between sleep mode and active mode in a strategy known as **service on-demand mode**.

The types of batteries used in these devices vary according to the device's requirements. For instance, smart tags often use button batteries and smart home devices may use alkaline dry batteries, while outdoor devices typically use rechargeable lithium batteries or 18650 batteries.

Connected by wire versus wireless

IoT end devices need to connect to the cloud to send or receive data. This connection can either be wired or wireless. In industrial IoT devices, such as those used in manufacturing plants or utility facilities, traditional wire connectivity such as fiber, ethernet, RS232/485 twisted pairs, and power cables are still commonly used alongside wireless connections. On the other hand, IoT end devices for home, campus, building, city, and rural areas widely use wireless connectivity.

The choice between wire or wireless connectivity depends on the specific application requirements, including factors such as mobility, roaming, reliability, latency, scalability, throughput, and cost.

For applications that demand high reliability, low latency, high throughput, and fixed installations, ethernet or fiber connections are the most cost-effective. These provide a stable and secure connection, ensuring uninterrupted data transmission. If mobility and roaming are essential, several options are currently available such as LTE CAT-4 or above, 5G, or Wi-Fi 6/6E/7. These technologies enable seamless mobility and uninterrupted service, making them suitable for applications such as AMRs, **Autonomous Guided Vehicles** (**AGVs**), and vehicle fleet management, where continuous communication is crucial.

However, if your applications do not mandate high reliability, low latency, and high throughput, but prioritize the total cost, massive scalability, and easy deployment, there are other options to consider. BLE, LoRaWAN, NB-IoT, **LTE CAT-M** (**LTE-M**), and **LTE CAT-1** are smart investments. These technologies offer low-cost solutions with wide coverage and low power consumption, making them suitable for applications where budget and scalability are the primary concerns. For example, BLE is widely equipped in smart home devices. LoRaWAN is popularly adopted for smart city applications. NB-IoT and LTE CAT-M are used for smart metering, and LTE CAT-1 is used for remote asset tracking.

The need for edge computing

Some IoT applications require their IoT end devices to support edge computing locally, while others do not. Edge computing is a local source of processing and storage for the data and computing needs of IoT devices, such as filtering the expected data through specific conditions, encrypting raw data payloads, executing protocol translations, triggering a local execution policy, and raising a prompt alarm.

There are several significant benefits that come from edge computing at IoT end devices, as follows:

- **Reducing latency**: By utilizing edge computing capabilities, IoT devices can process and analyze data locally. This significantly reduces the latency of communication between devices and enables faster response times. Real-time data insights can be generated at the edge, leading to increased operational efficiency and improved decision-making.

- **Increasing operational efficiency**: Edge computing allows for localized data processing and analysis. This means that IoT devices can perform tasks such as filtering and aggregating data, executing protocols, and making rapid decisions without relying on a centralized network. As a result, operational efficiency is improved, as devices can autonomously perform actions and respond to events in real time.

- **Saving network bandwidth**: By leveraging edge computing, IoT devices can offload some of the data processing tasks from the cloud. This reduces the amount of data that needs to be transmitted over the network, resulting in improved network bandwidth utilization. It also helps alleviate network congestion and ensures that critical data can be transmitted efficiently.

- **Continuing systems operation when offline**: One of the key advantages of edge computing is its ability to operate offline. In case of a network connection loss, IoT devices can still function and perform their designated tasks. This is particularly crucial in scenarios where uninterrupted operation is essential, such as in industrial environments or remote areas with limited connectivity.

Edge computing requires powerful MCU, as well as larger memory and storage sizes to handle local computing tasks efficiently. This, in turn, leads to an increase in the cost of hardware and the complexity of software at IoT end devices.

In addition to these four popular categories, there are many other criteria used to classify IoT devices within the market. Regardless of their category, the hardware architecture of IoT devices is typically similar, usually including an MCU, I/O interfaces, a sensor and/or actuator, a power supply, and so on.

Hardware architecture

The hardware architecture of IoT end devices is usually composed of a diverse array of components that work together seamlessly to enable their functionality. These components, such as MCUs, sensors, actuators, connectivity modules, and power management modules, collaborate to efficiently collect data from the objects that they are monitoring. They then process this captured data and transmit it to cloud platforms for in-depth analysis and informed decision-making. This collaborative effort ensures the smooth operation and effectiveness of IoT end devices throughout their life cycles.

The following figure depicts the typical hardware architecture of an IoT device:

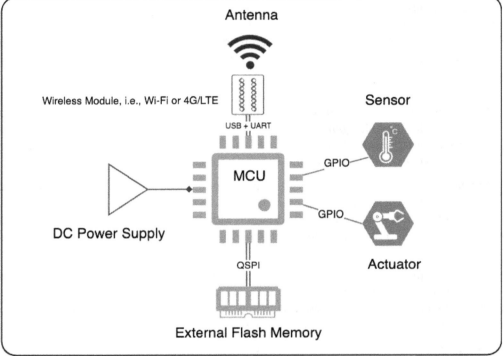

Figure 3.1 – IoT device hardware architecture

These components include the following:

- **MCUs**: MCUs act as the *brain* of the IoT device, orchestrating operations, processing data, and managing communication with peripherals and external networks. The integration of core processors, memory, I/O interfaces, and real-time clocks enables sophisticated computational capabilities and real-time operations.

- **Sensors and actuators**: Sensors serve as the device's senses, collecting environmental data or monitoring various parameters. This data becomes the foundation for decision-making and actions taken by the device. Actuators act upon the environment or the device itself based on processed data or commands received, creating a tangible output or change in the system.

- **Connectivity module**: This module provides the necessary interface for data exchange between the IoT device and external networks, including cloud platforms. It supports a wide range of communication protocols and technologies, ranging from local networks (for example, BLE, ZigBee, and Wi-Fi) to wide-area networks (for example, 4G/LTE, 5G, and LoRaWAN), ensuring flexibility and scalability in deployment.

- **Power management module**: This module ensures the efficient and reliable supply of power to the device, which is crucial for sustained operation, especially in remote or inaccessible locations. It incorporates features to optimize energy consumption. This is crucial for battery-operated devices, prolonging their operational lifespan and reducing maintenance needs.

- **External flash memory**: This provides additional storage capacity for program code, data logs, and configuration settings, enabling more complex applications and data retention. It enhances the device's capability to handle extensive data processing and logging, as well as to retain critical information during power cycles.

The effective combination of these components enables IoT devices to perform their tasks efficiently and effortlessly. In unison, they equip the device with the ability to collect, react, process, and communicate. This proves beneficial in various sectors such as smart homes, industrial automation, healthcare, and agriculture, fostering a more connected and intelligent environment.

By understanding the roles and significance of each component, developers and engineers can design IoT devices that function effectively, conserve energy, and cater to specific requirements and constraints. The upcoming section will provide additional details about the MCU, its peripherals and I/O interfaces, sensors, and actuators. *Chapter 4* will specifically focus on the wireless connectivity utilized by IoT end devices.

MCUs

An MCU is the foundation of embedded systems. These compact yet potent devices combine core processors, built-in memory, and a vast range of peripherals into one chip. This combination leads to highly efficient power usage, space-saving, and cost-effectiveness, all while maintaining robust processing capabilities.

Roles

In embedded systems, the MCU serves as the central processing unit. It is responsible for executing and controlling a diverse array of tasks and functions. With its prowess in handling complex algorithms and interfacing with various external components, the MCU forms the backbone of the system, facilitating seamless communication and coordination among different hardware and software components. It is this orchestration that enhances the system's overall functionality and reliability, making the MCU an indispensable component in a myriad of electronic devices and applications.

Key features

MCUs are characterized by their purpose-driven design, offering ease of programming and customization to meet the specific demands of varying applications and projects. This versatility allows developers to tailor the MCU's behavior, optimizing its functionality and performance. For instance, MCUs crafted for IoT applications are celebrated for their low power consumption, an essential feature for battery-dependent or energy-sensitive applications. Moreover, the integration of multiple components into a single chip underscores the MCU's popularity in the embedded systems domain.

Critical components

MCU usually includes multiple key components like processor, memory, peripherals, interfaces, etc. The upcoming sections will cover information about these.

Core processor

- **Architecture**: Including ARM, AVR, MIPS, and RISC-V, each offering unique strengths
- **Processing clock speed**: Measured in MHz or GHz, this determines the rate at which instructions are executed

Built-in memory

- **Random Access Memory (RAM)**: For transient data storage during operations
- **Read-Only Memory (ROM)**: Houses the bootloader and permanent software

Peripherals and interfaces

- **General Purpose Input/Output (GPIO)** : Programmable pins for interfacing with sensors, actuators, LEDs, and so on
- **Specialized I/O**: Provides input and output interfaces such as **Universal Asynchronous Receiver/Transmitter (UART)** for serial communication, **Serial Peripheral Interface (SPI)**, and **Inter-Integrated Circuit (I2C)** for interfacing with other chips
- **Pulse Width Modulation (PWM)**: Utilized in controlling motors, LEDs, and so on

Connectivity module

- **Wire**: Includes Ethernet, **Universal Serial Bus (USB)** , RS232/485, and so on
- **Wireless**: Spans Wi-Fi, BLE, ZigBee, Thread, LoRaWAN, 4G/LTE and 5G, and so on

Power management

- **Voltage range**: Dictates the required power supply
- **Low power modes**: Critical for conserving energy in battery-powered devices

Embedded operating system

- **Real-Time Operating System (RTOS)**: Supports complex demands requiring structured task management; examples include FreeRTOS, Azure RTOS, VxWorks, Zephyr, and so on
- **Bare-metal OS**: Enables direct control over the hardware for simpler systems.

Off-the-shelf MCU

The following list contains the popular semi-conductor vendors' commodity MCU products that are widely adopted for building IoT end devices. These products offer a wide range of features and capabilities, making them ideal choices for developers and engineers in the IoT industry. Let us take a closer look at each of these MCU series:

- **Microchip Technology PIC series**: Known for their reliability and versatility, the PIC series from Microchip Technology is a popular choice among IoT device builders.

- **STMicroelectronics STM32 series**: The STM32 series from STMicroelectronics is highly regarded for its performance and power efficiency, making it a preferred option for IoT applications.

- **Texas Instruments C2000/C3000 series**: Texas Instruments' C2000/C3000 series of MCUs are known for their real-time control capabilities, making them suitable for a wide range of IoT projects.

- **Nordic nRF51/nRF52 series**: The nRF51/nRF52 series from Nordic Semiconductor offers excellent wireless connectivity options, making it a top choice for IoT devices that require reliable communication.

- **Silicon Labs' EFM32 Gecko series**: The EFM32 Gecko series from Silicon Labs is known for its low power consumption and energy efficiency, making it ideal for battery-powered IoT devices.

- **Particle Photon series**: The Particle Photon series is a popular choice among IoT developers due to its simplicity and ease of use, offering seamless cloud connectivity for IoT applications.

These MCU series provide a solid foundation for building IoT end devices, enabling developers to create innovative and efficient solutions for the rapidly growing IoT market.

DIY-friendly MCU

For beginners embarking on their first IoT project, it is highly recommended to explore the DIY-friendly MCU options provided by Raspberry Pi, Arduino Families, and ESP8266/ESP32.

These platforms offer a range of user-friendly features and robust ecosystems, making them ideal for beginners to immerse themselves in the exciting world of IoT. Raspberry Pi Pico boasts a compact size and powerful performance. Arduino Families provides a wealth of options, as well as intuitive interfaces and a large community of ecosystem developers. ESP8266/ESP32, built on RISC-V architecture, enables beginners to explore various aspects of IoT development.

Raspberry Pi

Raspberry Pi is a popular open source hardware family that is widely used in a variety of IoT engineering projects. Its versatility, affordability, powerful processing capabilities, and extensive connectivity options make it a top choice for hobbyists, students, and professionals. Raspberry Pi has revolutionized the world of DIY electronics, enabling the creation of home automation systems, robotics projects, and more. No matter your level of experience, Raspberry Pi provides an accessible platform to explore electronics and programming.

The following figure depicts the product series in its family.

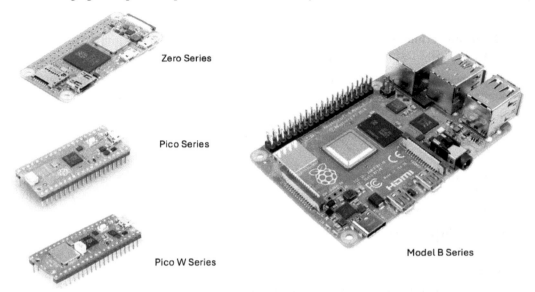

Figure 3.2 – The Raspberry Pi family

These series are designed to meet diverse needs and project scales:

- **Raspberry Pi 5**: This is the latest iteration in the Raspberry Pi series of single-board computers (quad-core ARM Cortex-A76 processor), designed for both educational and practical applications.

- **Raspberry Pi 4 Model B**: This is equipped with a Quad-core ARM Cortex-A72 processor, known for its balance of performance, connectivity, and affordability. It is ideal for those needing a full-featured computer for coding, content creation, or general use.

- **Raspberry Pi Zero**: This has a a single-core ARM11 processor (BCM2835) embedded. It prioritizes compactness and efficiency. It is perfect for space-constrained projects and applications where minimal power consumption is important.

- **Raspberry Pi Pico**: This was built on a Dual-core ARM Cortex-M0+ processor, offering a cost-effective solution for embedded systems, IoT devices, and prototyping.

- **Raspberry Pi Pico W**: This introduces Wi-Fi and BLE connectivity on Pico.

You can find more details at https://www.raspberrypi.com/.

Arduino

Arduino is a widely embraced open source hardware platform that is popular for IoT end-device development. Arduino has changed and grown over time, resulting in diverse types such as Classic, Nano, MKR, and Mega. Each type has its own specific features. Because of this, Arduino is now a flexible and powerful platform for IoT development and more. Whether you are new or experienced, there is bound to be an Arduino board that fits your needs and helps you bring your ideas to life.

The following figure depicts the product series in its family.

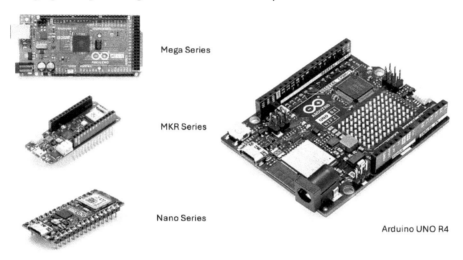

Figure 3.3 – The Arduino family

These series are designed to accommodate diverse needs and various project scales:

- **The Arduino Classic family**: This includes the popular Arduino UNO with the ATmega328P core processor (Single-core 8-bit AVR), as well as other classic models such as the Leonardo & Micro. It is the base option of the Arduino project.

- **The Arduino Mega family**: This is a series of boards designed for projects that require significant computing power and many GPIO pins. The Arduino Mega 2560 model is equipped with the ATmega2560 core processor, an 8-bit AVR, while the Arduino Due is built with the Atmel SAM3X8E ARM Cortex-M3 CPU, a 32-bit ARM Cortex.

- **The Arduino Nano family**: This is a collection of small boards suitable for IoT projects. The family includes the basic Nano (ATmega328P, 8-bit AVR), Nano Every (ATmega4809, 8-bit AVR), Nano 33 IoT (32-bit ARM Cortex), Nano 33 BLE, Nano 33 BLE Sense (32-bit ARM Cortex), and Nano RP2040 (Dual-core ARM Cortex M0+). These boards have Bluetooth and Wi-Fi modules, as well as built-in sensors such as temperature, humidity, pressure, gesture, and microphone sensors.

- **Arduino MKR Family**: This is a series of boards, shields, and carriers that can be combined to create amazing projects without the need for additional circuitry. Each board is equipped with a radio module (except for MKR Zero) that enables Wi-Fi, Bluetooth, LoRa, SigFox, and NB-IoT communication. All boards in the MKR family are based on the 32-bit ARM Cortex-M0+, a low-power processor, and come with a crypto chip for secure communication.

You can find more details at `https://www.arduino.cc/en/hardware`.

ESP8266 and ESP32

ESP8266 and ESP32 MCUs are popular MCUs based on the RISC-V architecture. They are known for their versatility, reliability, and wide range of features. These MCUs have gained significant attention and adoption in industries such as IoT, home automation, and robotics.

Reduced Instruction Set Computer-V (RISC-V) is an open source **Instruction Set Architecture (ISA)** that is becoming increasingly popular in computer architecture. It offers a simple and modular design, which makes it easier to understand and implement compared to other complex architectures. Being open source, RISC-V provides greater flexibility and customization options, allowing developers to adapt the architecture to their specific needs.

ESP8266 and ESP32 are Espressif products. You can find more details at `https://www.espressif.com/en/products/modules`. The following figure shows ESP8266 and ESP32 products.

ESP8266 Series   ESP32 Series

Figure 3.4 – The ESP8266 and ESP32 families

ESP8266 and ESP32 are designed to meet various needs and accommodate projects of different scales:

- **ESP8266**: This product, which was introduced in 2014, brought about a momentous change in the world of MCU. It made Wi-Fi connectivity affordable and accessible, allowing developers to easily connect their devices to the internet. The ESP8266 quickly gained popularity and became a top choice for many IoT projects.

- **ESP32**: This product was released in 2016. Built upon the success of the ESP8266, it took the capabilities of its predecessor to the next level, offering not only Wi-Fi connectivity but also Bluetooth, dual-core processing, and more GPIO pins. The ESP32 revolutionized the industry, empowering developers to create even more advanced and feature-rich IoT applications.

Developers can use the ESP8266 and ESP32 series to access a variety of development tools, libraries, and resources. The Espressif development ecosystem offers detailed documentation, sample codes, and tutorials to help developers get started and create their projects. They can also use Arduino IDE and Microsoft Visual Studio Code IDE with the PlatformIO extension for development tools.

Here is a comparison table outlining the key specifications of the ESP32 and its variants:

Specification	ESP32	ESP32-S2	ESP32-S3	ESP32-C3	ESP32-C6	ESP32-H2
Launch date	2016	2019	2020	2020	2023	2023
Processor	Dual-core 32-bit Xtensa LX6	Single-core 32-bit Xtensa LX7	Dual-core 32-bit Xtensa LX7	Single-core RISC-V 32-bit	Single-core RISC-V 32-bit	Single-core RISC-V 32-bit
CPU frequency	240 MHz	240 MHz	240 MHz	160 MHz	160 MHz	96 MHz
Wi-Fi, 2.4GHz	Wi-Fi 4 (802.11 b/g/n)	Wi-Fi 4 (802.11 b/g/n)	Wi-Fi 4 (802.11 b/g/n)	Wi-Fi 4 (802.11 b/g/n)	Wi-Fi 6 (802.11 ax)	Not supported
Bluetooth	Bluetooth 4.2 (LE)	Not Supported	Bluetooth 5.2 (LE), Bluetooth Mesh	Bluetooth 5.2 (LE), Bluetooth Mesh	Bluetooth 5.3 (LE), Bluetooth Mesh	Bluetooth 5.3 (LE), Bluetooth Mesh
802.15.4	None	None	None	None	ZigBee 3.0, Thread 1.3	ZigBee 3.0, Thread 1.3
On-chip RAM	520 KB	320 KB	512 KB	400 KB	512 KB	320 KB
On-chip ROM	448KB	128KB	384 KB	384 KB	320 KB	128 KB
Programmable GPIO pins	34	43	38	22	22	19
SPI	4	4	4	3	2	3
UART interface	3	2	3	2	2	2
I2C interface	2	2	2	1	1	2
I2S interface	2	1	2	1	1	2

Specification	ESP32	ESP32-S2	ESP32-S3	ESP32-C3	ESP32-C6	ESP32-H2
Analog-to-Digital Converter (ADC) channel	18	20	20	5	7	5
Built-in temperature sensor	1	1	1	1	1	1

Table 3.2 – The key specifications of the ESP32 series

> **Note**
>
> Please note that we will use the ESP32-C3 in the practice sessions starting from *Chapter 11*.

An MCU combines core processors with various peripherals and interfaces into a single module. The next section will discuss the common peripherals and interfaces typically found inside MCUs.

Peripherals and interfaces

MCU features various peripherals and interfaces, including GPIO, SPI, I2C, UART, USB, **Secure Digital Input Output (SDIO)**, ADC, **Digital-to-Analog Converter (DAC)**, PWM, and **Joint Test Action Group (JTAG)**, among others.

GPIO

GPIO is a set of programmable pins that can be configured as input or output ports to handle digital signals from sensors and actuators. When configured as an input port, it can receive ON and OFF signals from switches, or digital readings from sensors, and transmit them to the MCU. Conversely, when configured as an output port, it can execute external operations according to MCU instructions. For example, it can blink an LED or generate drive signals for a motor.

GPIO pins are commonly used to connect sensors and actuators.

SPI

SPI port is a widely used synchronous serial communication bus specifically designed for point-to-point, short-distance, and high-data-rate communication, particularly within embedded systems. Its main function is to facilitate data transfer between an MCU and peripheral devices, such as **Secure Digital (SD)** cards, LCD displays, sensors, and other peripherals.

SPI port typically uses four wires for data communication. These wires are as follows:

- **Serial Clock (SCLK)**: This is the clock signal provided by the master device. It synchronizes the data transmission between the master and slave devices.

- **Master Out Slave In (MOSI)**: This is the data line used for sending data from the master to the slave device.

- **Master in Slave Out (MISO)**: This is the data line for sending data from the slave back to the master.

- **Slave Select (SS)**: This is a control line used by the master to select and control individual slave devices. When multiple slave devices are connected to an SPI bus, each one will have its own SS line.

I2C

I2C is a widely used communication bus in the field of electronics. It is a multi-master, multi-slave, packet-switched, single-ended, and serial communication bus. The I2C bus is primarily designed for connecting low-speed devices such as EEPROMs, sensors, and other MCUs. It provides a simple and efficient way for these devices to communicate with each other.

The I2C bus allows multiple devices to be connected, with each device having a unique address. This enables easy integration and communication between different components in a system. The bus operates in a master-slave configuration, where the master device initiates communication and controls the data transfer between the devices.

I2C port typically uses two wires for data communication. These two wires are as follows:

- **Serial Data Line (SDA)**: This is used for transferring data between the devices on the I2C bus.

- **Serial Clock Line (SCL)**: This is the clock signal used to synchronize data transfer between devices on the I2C bus.

UART

UART is a hardware component that facilitates asynchronous serial communication. Asynchronous communication is not synchronized between parties. This means that the sender does not have to wait for the receiver or a broker to determine when to send or not to send.

UART is commonly used for communication between a PC and MCU. It is also utilized in RF wireless communication, sensors (such as GPS sensors), and old dial-up modems.

UART port typically requires a minimum of two wires for basic communication. These two wires are:

- **Transmit (TX)**: This wire is used for transmitting data from the MCU to another device.

- **Receive (RX)**: This wire is used for receiving data from another device to the MCU.

For basic UART communication, only these two wires are essential. However, in more complex or specific applications, additional wires might be used for hardware flow control or signaling purposes:

- **Clear to Send (CTS)**: Used for flow control, indicating that the MCU is ready to receive data

- **Request to Send (RTS)**: Also used for flow control, indicating that the MCU wants to send data

- **Ground (GND)**: A common ground connection is also typically necessary between communicating devices

In a simple UART setup, particularly in basic point-to-point communication scenarios, just the TX and RX lines, along with a common ground, are sufficient.

USB

The **USB** port is a widely used interface that facilitates data transfer and power supply between computers and electronic devices. It serves multiple purposes, including programming the MCU, enabling serial communication, and connecting peripherals such as keyboards and mice. The USB port provides a convenient and reliable method for establishing a connection between a computer and various electronic devices, enhancing their functionality, and enabling seamless data exchange.

SDIO

SDIO is an interface used in MCU and other devices for communication with SD cards and other types of IO devices. It is an extension of the standard SD card protocol, allowing devices to not only transfer storage data but also use peripheral functions such as Bluetooth, GPS, and others.

ADCs

ADCs are electronic devices that play a crucial role in converting analog signals, such as those generated by temperature sensors, into digital values that can be processed by the MCU.

By doing so, ADC enables the MCU to accurately interpret and utilize the information provided by various analog sensors, including but not limited to thermistors and potentiometers. These sensors, with their ability to measure temperature and provide variable resistance, respectively, are widely used in a variety of applications, ranging from industrial control systems to consumer electronics. The conversion process carried out by ADC ensures that the MCU can effectively analyze and respond to these analog inputs, enabling precise and reliable measurements, control, and decision-making.

DACs

DACs are electronic devices that are used to convert digital values from the MCU into analog signals. This conversion process is essential for various applications, including generating **Analog Outputs (AOs)** for audio and creating smooth waveforms.

By utilizing DACs, the digital information stored in the MCU can be transformed into continuous and varying analog signals. This enables the accurate representation of audio signals, allowing MCUs to produce high-quality sound outputs. Additionally, DACs play a crucial role in synthesizers, music production, and other audio-related fields.

PWM outputs

PWM outputs are a versatile feature that allows for the generation of a digital signal with varying pulse width, which effectively simulates an analog signal. This functionality opens up a wide range of possibilities for controlling various components and devices.

One of the key applications of PWM outputs is in controlling the speed of motors. By adjusting the pulse width of the digital signal, the rotational speed of motors can be finely tuned, providing precise control over their movements.

Additionally, PWM outputs are commonly used in driving servo motors. By varying the pulse width of the digital signal, the angular position of the servo motor's shaft can be accurately controlled, enabling precise positioning in robotics and automation systems.

JTAG

JTAG is a standardized interface for debugging and testing **Integrated Circuits** (**ICs**), including MCUs such as the ESP32. It was originally designed for testing printed circuit boards but has since evolved into a widely used tool for in-circuit debugging.

Timers

Timers are an essential tool for keeping track of time. They operate by maintaining a counter that incrementally advances at a predetermined rate. This functionality can be utilized for various purposes, such as the following:

*   Monitoring the duration of events or processes

*   Scheduling and coordinating tasks

*   Implementing time-based actions or triggers

*   Measuring intervals or time lapses

*   Creating countdowns or alarms

Real-time clock

Real-Time Clock (**RTC**) is a peripheral component that is designed to operate continuously, even during low-power sleep modes, to accurately keep track of time.

The RTC provides two essential timing functions: the RTC and the **Periodic Interrupt Timer** (**PIT**). The RTC keeps track of the current time and date, while the PIT allows the device to be woken up from sleep modes or interrupted at regular intervals. With these functions, the RTC not only ensures accurate timekeeping but also enables the device to effectively manage power consumption and perform tasks at specific time intervals.

These peripherals and interfaces, which include a diverse range of devices and software, are actively engaged in the interaction with various sensors and actuators. This interaction facilitates the collection of a wide array of data, from environmental factors to user inputs. Furthermore, it also enables the execution of specific actions based on this data. This sophisticated interplay between peripherals, interfaces, sensors, and actuators forms the foundation of our data-driven world, allowing for enhanced automation and decision-making capacities. Now, let's look at another important topic.

Sensors and actuators

In the vast landscape of IoT, end devices are equipped with an array of sensors and actuators, forming the cornerstone for dynamic interaction with the physical world. These components not only perceive the nuances of their environment but also act upon it, paving the way for intelligent and responsive IoT applications.

Sensors

Sensors in IoT devices are used to capture and send environmental data to the MCU. They can detect various parameters, such as temperature, humidity, light, and motion. Sensors provide valuable information about the environment, helping with decision-making and adaptive actions.

Many sensors used in IoT devices are **Micro-Electro-Mechanical Systems** (**MEMS**) devices. MEMS technology, which enables the miniaturization of mechanical and electro-mechanical elements to a small size, has had a significant impact on sensor technology. By combining mechanical and electro-mechanical elements with electronic components on a silicon substrate, MEMS sensors offer many advantages. These sensors are widely used in IoT devices and consumer electronics because they are small, consume little power, and have excellent performance. As a result, MEMS technology has revolutionized sensor technology, allowing for the development of smaller, more efficient, and higher-performance devices in the field of IoT and consumer electronics.

Here are some examples of sensors:

- **Temperature sensor**: Measures ambient temperature, with variants such as thermistors and thermocouples providing precision
- **Humidity sensor**: Gauges moisture content; vital for climate control systems
- **Pressure sensor**: Monitors atmospheric or water pressure; essential in weather forecasting and industrial applications

- **Proximity sensor**: Detects nearby objects sans physical contact using ultrasonic or infrared technologies

- **Light sensor**: Measures luminosity, employing photodiodes or LDRs to adapt lighting or monitor environmental conditions

- **Motion and occupancy sensor**: Detects movement, leveraging technologies such as PIR for security and automation

- **Gas sensor**: Identifies various gases; crucial for safety and environmental monitoring

- **Air quality sensor**: Gauges air pollutants; vital for health and environmental quality assessments

- **Sound sensor**: Monitors acoustic levels, employing microphones or piezoelectric sensors for noise measurement or user interaction

- **Water quality sensor**: Assesses water parameters; pivotal in ensuring water safety and quality

- **Gyroscopes and accelerometers**: Track orientation and motion; integral in navigation and motion-sensitive devices

- **Heart rate sensor**: Monitors cardiac rhythms; essential in health and fitness devices

Actuators

Conversely, actuators translate the digital data gleaned from sensors into tangible actions. They are the muscles of IoT devices, enabling them to exert a physical influence on their surroundings. Whether by modulating light intensity, regulating a motor's speed, or actuating valves, actuators respond to sensor inputs or remote commands, fostering a responsive and adaptive environment.

Here are some examples of actuators:

- **Motors (DC, Stepper, Servo)**: Drive mechanical systems, facilitating motion and precise control

- **Relays**: Serve as electrically operated switches, managing larger loads, commonly in home automation

- **Solenoids**: Electromagnetic actuators, employed in locking systems and fluid control

- **LEDs and displays**: Provide visual feedback or interfaces; used as status indicators or information displays

- **Buzzers and speakers**: Generate audio signals for alerts or user interaction; integral in alarm systems

- **Heaters and coolers**: Manage temperature for comfort or equipment safety

- **Pumps**: Propel liquids; utilized in applications such as automated irrigation or smart aquariums

- **Servo mechanisms**: Offer meticulous control over position; essential in robotics and precision applications

Common pins on sensors

Integrating sensors with MCUs involves interfacing through various pins, each serving a specific function. Proper understanding and connection of these pins is vital for accurate data collection and device performance. Common pin definitions include the following:

- **Voltage Common Collector** (**VCC**): This pin is used to supply power to the sensor. It is usually connected to the power source and its voltage level must match the sensor's requirements (commonly 3.3V or 5V).

- **GND**: The ground pin provides a common reference point for the power supply and signal levels. It is essential for completing the circuit and stabilizing the sensor's operation.

- **Data Pin** (**DATA**): This pin transmits the sensor's data to the MCU. The nature of data transmission can vary – it might be analog or digital and the protocol can be I2C, SPI, UART, or a proprietary one.

- **SCL and SDA**: These pins are specifically used in I2C communication. SCL carries the clock signal and SDA carries the data. Both lines are bidirectional, allowing multiple sensors to be connected to the same bus.

- **MISO, MOSI, and SCK**: Used in SPI communication, these pins facilitate full-duplex data transfer. SCK is the clock line, MOSI sends data from the MCU to the sensor, and MISO sends data from the sensor to the MCU.

- **TX and RX**: In UART communication, TX and RX are used for serial data transfer. The TX of the MCU connects to the RX of the sensor and vice versa.

- **AO** and **Digital Output** (**DO**): Some sensors offer both analog and digital outputs. AO provides a variable voltage level proportional to the measured parameter, while DO provides a binary output, indicating whether the parameter is above or below a certain threshold.

- **INT** (**Interrupt**): The interrupt pin allows the sensor to notify the MCU of certain events (such as threshold crossing or data readiness) without the need for continuous polling.

As an example, let us look at the DHT11 temperature and humidity sensor pinout. A commonly used temperature and humidity sensor, it typically has the following pins:

- **VCC**: Power supply (between 3.3 and 5V)

- **DATA**: Outputs both temperature and humidity data

- **GND**: Ground

For the DHT11, a single DATA transmits both temperature and humidity readings in a serial digital form. It supports a one-wire protocol that uses a timing-based communication scheme. The MCU must use specific timing to read this data correctly.

Understanding sensor specifications

Datasheets are a valuable resource for understanding what sensors and actuators can do and what their limitations are. They might seem overwhelming at first, but learning how to find useful information in datasheets is an important skill. Many sensor and actuator modules are designed to work easily with popular development boards such as Arduino, Raspberry Pi, and ESP8266/ESP32. These modules often come with libraries and sample code that can help beginners get started quickly. Some sensors might require calibration to make sure that they give accurate readings. Make sure that you know how to calibrate your sensors and do it in the same environment where you will use the sensor.

The following list contains general advice for beginners to understand sensor specifications before building their innovation prototype:

- **Sensitivity**: This indicates how much the sensor's output changes when the measured quantity changes. A more sensitive sensor can detect smaller changes in the measured parameter. For example, a high-sensitivity grade temperature sensor can accurately detect even a slight change in temperature, making it ideal for applications where precise temperature monitoring is required.

- **Range**: The range of a sensor refers to the minimum and maximum values of the measured parameter that the sensor can accurately detect. For instance, a pressure sensor with a wide range can accurately measure both low and high pressures, making it suitable for various industrial applications.

- **Resolution**: This defines the smallest change in the measured quantity that the sensor can detect. A sensor with a higher resolution can detect smaller changes. For instance, a high-resolution pH sensor can detect subtle changes in acidity levels, making it beneficial for precise pH monitoring in scientific experiments.

- **Accuracy**: This refers to how close the sensor's readings are to the actual value. It is important to consider both the accuracy and the precision (repeatability) of the sensor. For example, a highly accurate weight sensor can provide precise measurements with minimal deviation from the true weight, ensuring reliable data for applications such as industrial weighing scales.

- **Response time**: This refers to the time that a sensor takes to respond to a change in the measured quantity. A faster response time means that the sensor can accurately track rapid changes. For instance, a high-speed motion sensor can quickly detect and respond to fast movements, making it suitable for applications such as sport tracking or security systems.

- **Power consumption**: This refers to the amount of power consumed by the sensor during operation. Lower power consumption is desirable as it prolongs battery life and reduces energy costs. For example, an energy-efficient light sensor can effectively detect changes in ambient light levels while consuming minimal power, making it suitable for battery-powered devices such as smartphones or smart watches.

Summary

In this chapter, we learned about IoT end devices used in IoT systems. We discussed their types and hardware architecture, as well as the importance of MCUs. We also explored the important parts and connections that allow these devices to interact with their surroundings. Additionally, we talked about sensors and actuators that gather data and enable actions in the real world.

As we move on to the next chapter, we will build on this knowledge by exploring how wireless data communication works in the air. Wireless connectivity is what connects individual devices into an IoT network that can transmit data and commands to fully utilize the potential of IoT. We will look at different wireless connectivity technologies in detail, including Wi-Fi, BLE, LTE-M, and NB-IOT in the domain of 4G/LTE and 5G.

4

Wireless Connectivity, the Nervous Pathway to Delivering IoT Data

In the architecture of IoT, wireless connectivity functions as a conduction subsystem, like the nervous pathways within the human body that transmit *biological electrical* signals (sensor data or actuator command) between neural cells (IoT end devices) and the brain (backend cloud platform).

In this chapter, you will learn the fundamental knowledge points about wireless data communication that beginners should know when building their IoT innovations. This knowledge will provide you with a solid foundation to understand the intricacies of wireless data transmission, receiving, and signal processing.

By exploring the details of popular wireless technologies such as BLE, Wi-Fi, LTE-M, and NB-IOT in the domains of 4G/LTE and 5G, you will gain a comprehensive understanding of their capabilities and limitations.

By the end of this chapter, you will have acquired a wealth of knowledge that will empower you to confidently navigate the realm of wireless data communication in the context of IoT innovation.

In the engineering projects from *Chapter 11* onward, with the assistance of ChatGPT, you will practice Wi-Fi client stack coding on ESP32 MCU to access the internet through your home Wi-Fi network.

This chapter covers the following topics:

- 10 knowledge points about wireless data communication
- BLE
- Wi-Fi
- 4G/LTE and 5G

10 knowledge points about wireless data communication

In *Chapter 2*, we discussed diverse IoT networks covered by BLE, Wi-Fi, LoRaWAN, 4G/LTE, 5G, and LEO. They all fall under the domain of wireless data communication. Understanding how wireless data communication works from the Physical Layer in detail can be challenging and daunting, especially for people without a strong professional background.

Nevertheless, for beginners who strive to apply their IoT innovation to business, it is necessary to spend time grasping the intricacies of the low layers of wireless technology, such as the Physical Layer and Data Link Layer. Knowing the basics of how wireless data communication works in the field can be helpful for your innovation development process and field pilot.

To assist you on this journey, we'll cover 10 important technical facts about wireless data communication that beginners should be aware of. These facts will serve as a foundation for your understanding and enable you to delve deeper into the subject matter.

OSI model

The architecture model of wireless data communication technology follows a widely accepted industry standard called the **Open Systems Interconnection (OSI)** model. OSI is a framework that's used to understand and standardize the functions of a data communication system, regardless of whether it is wired or wireless. The OSI model was developed by the **International Organization for Standardization (ISO)** in the 1970s. It was designed to organize and streamline the different functions of data communication effectively. The OSI model divides the data communication end-to-end functions into seven layers, each with a specific purpose. The following table shows the seven-layer model and how each layer contributes to the data communication process:

Layers	Function Description
7, Application	Interacts with software applications that act as a data communicating component. It facilitates services such as email, file transfer, web browsing, instant messaging, voice and video telephony, and video streaming.
6, Presentation	Translates data between the format the network requires and the format the computer expects. It handles data encryption and decryption, as well as data compression, such as **Secure Sockets Layer (SSL)** encryption.
5, Session	Manages the sessions or connections between applications. It establishes, manages, and terminates connections between two or more devices.
4, Transport	Ensures complete data transfer and controls the reliability of a given link through flow control, segmentation/desegmentation, and error control, such as **Transmission Control Protocol (TCP)** and **User Datagram Protocol (UDP)**.
3, Network	Manages device addressing, tracks the location of devices on the network, and determines the best way to move data. It is responsible for routing and forwarding packets across network boundaries, such as **Internet Protocol (IP)** addresses.

Layers	Function Description
2, Data Link	Comprises two sublayers: **Logical Link Control** (LLC) and **Media Access Control** (**MAC**). It is responsible for node-to-node data transfer and error detection and correction in the physical layer, such as MAC addresses in Ethernet and Wi-Fi.
1, Physical	Deals with the physical connection between devices and the transmission and reception of raw bit streams over a physical medium (such as cables, fiber optics, or wireless).

Table 4.1 – The OSI model's 7 layers

It is important to understand that when it comes to different wireless communication technologies, there may not be a big difference between Layer 3 (Network) and Layer 7 (Application). However, it is worth noting that there can be significant differences in functionality and characteristics at Layer 1 (Physical) and Layer 2 (Data Link). These differences at the lower layers can have a significant impact on the overall performance and efficiency of the wireless communication system. So, it is necessary to carefully analyze and evaluate Layer 1 and Layer 2 as they play a critical role in shaping the behavior and capabilities of the wireless network.

Physical layer

The physical layer is the most important in the OSI model, especially for wireless communication technologies such as BLE, Wi-Fi, and 4G/LTE. This layer is responsible for sending and receiving the raw bitstream over the physical medium, which is done using electromagnetic waves in wireless technologies. The transmission process involves a few key steps: converting digital bits into analog signals using DAC, preparing these signals for transmission through proper modulation, and sending them using the antenna. On the receiving side, these steps are done in reverse order: the antenna receives the electromagnetic waves or signals, which are then demodulated and converted back to digital signals using ADC. We'll cover these in more detail in the *Signal processing* section.

Data link layer

The data link layer is an important part of the data communication process, especially in wireless networks. It is located above the physical layer in the OSI model. This layer has two sub-layers: the **Logical Link Control** (LLC) Layer and the **Media Access Control** (MAC) layer. Each sub-layer has specific functions to ensure effective and reliable communication. Some of the tasks it handles are error detection and correction, frame synchronization, and controlling access to the physical medium. In wireless communications, this layer also includes channel encoding/decoding and interleaving/de-interleaving, which are important for reliable data transmission and work closely with the physical layer.

The approach to medium access varies across different wireless technologies. For instance, Wi-Fi operates on a contention-based model, where clients compete with each other to access the spectrum for data transmission. This process is typically managed using protocols such as **Carrier-Sense Multiple Access with Collision Avoidance** (**CSMA/CA**). Conversely, in 4G/LTE networks, the access to the spectrum is more structured and is coordinated by the base station, known as the 4G/LTE eNodeB. In this scenario, the eNodeB allocates specific time and spectrum resources to each 4G/LTE client, facilitating organized and scheduled access to the spectrum for efficient data transmission.

Signal processing

The signal processing of wireless data communication is a complex and intricate process that involves numerous steps. Data transmission begins with encoding and modulating the information to be sent. This ensures that the data is converted into a suitable format for reliable wireless transmission. Next, the modulated signal is transmitted through the air using various techniques such as frequency hopping or spread spectrum. This helps ensure that the signal is robust and can withstand interference or noise. Once the signal reaches the receiver, it undergoes demodulation to extract the original information. Finally, the decoded data is processed and interpreted by the receiving device, allowing the recipient to access and utilize the transmitted information.

The following flow chart is a basic diagram representing the signal process for wireless data communication:

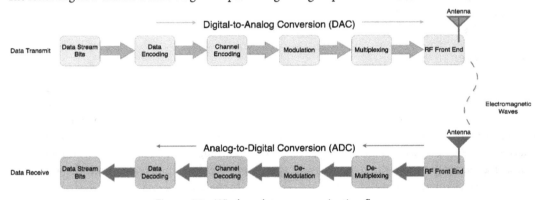

Figure 4.1 – Wireless data communication flow

Electromagnetic waves

According to Maxwell's equations, wireless data is propagated in the form of electromagnetic waves in the air. Maxwell's equations are the basis of classical electromagnetism, classical optics, and electric and magnetic circuits. These equations give a mathematical model for electric, optical, and radio technologies, such as power generation, electric motors, wireless communication, lenses, radar, and more. They explain how electric and magnetic fields are created by charges, currents, and changes in the fields.

Electromagnetic waves are formed by two oscillating fields: an electric field (E) and a magnetic field (B). The electric and magnetic fields oscillate in phase, meaning there is no phase shift between the waveforms of the electric and magnetic fields, so the peaks and troughs of these fields are perfectly aligned, resulting in synchronized oscillations. Secondly, the electric and magnetic fields are orthogonal to each other, as is their direction of propagation, forming 90° angles with one another.

The following is a 3D view of electromagnetic waves in the air:

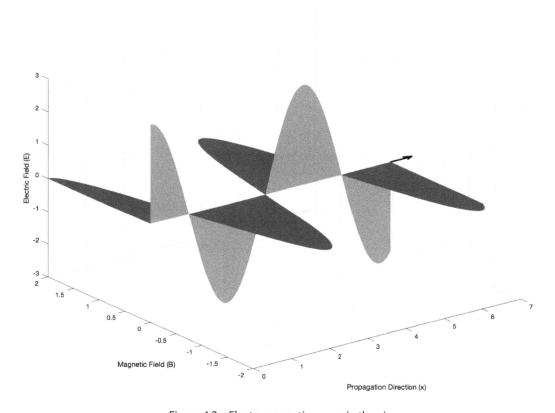

Figure 4.2 – Electromagnetic waves in the air

It is important to know that acoustic waves and electromagnetic waves are different. Acoustic waves, also called sound waves, are created when particles vibrate in a medium such as air or water. Electromagnetic waves are made when electric and magnetic fields interact, and they do not need a medium to travel through.

In our daily lives, sunlight consists of natural electromagnetic waves. Besides the commonly known wireless communication technologies, such as Wi-Fi, Bluetooth, and cellular networks, electromagnetic waves are also utilized in radio and TV broadcasting, microwave ovens, remote controls, GPS, military and civil radar systems, home illumination light, and medical X-rays, among other applications.

Frequency and wavelength

Frequency and wavelength play a vital role in determining how electromagnetic waves propagate and interact with the environment. Understanding the relationship between frequency and wavelength is essential for comprehending how wireless signals are transmitted, received, and manipulated.

Frequency

Frequency is a unit that's used to measure how often the same waves (regardless of electromagnetic waves or sound waves) are repeated within a given timeframe, typically expressed in cycles per second or **hertz (Hz)**. The frequency of sound waves that people can hear spans from about 20 Hz to 20,000 Hz (20 kHz). The frequency that's allowed for Wi-Fi is 2.4GHz, 5GHz, and 6GHz (Wi-Fi 6E).

The following figure shows the frequency waveform of 2.4 GHz versus 5 GHz in a-1 nanosecond (**10^-9** seconds) period:

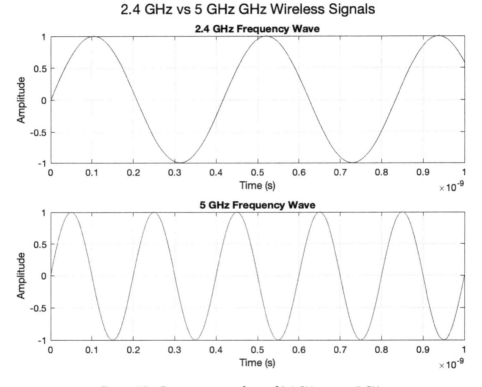

Figure 4.3 – Frequency waveform of 2.4 GHz versus 5 GHz

Wavelength

Wavelength is the distance between successive peaks (or any identical points) in a wave, measured in **meters (m)**. The relationship between frequency and wavelength in electromagnetic waves is inverse and governed by a simple equation:

Speed of Light = Frequency × Wavelength

Here, the speed of light (in a vacuum) is approximately 3×10^8 **meters per second (m/s)**.

From this equation, you can see that as the frequency increases, the wavelength decreases, and vice versa. The higher the frequency of a wave, the shorter its wavelength.

For example, visible sunlight ranges from 400 to 700 **nanometers** (**nm**) in wavelength. In terms of frequency, this corresponds to about 430 **terahertz** (**THz**) to 750 THz. 5G **millimeter wave** (**mmWave**) in cellular networks refers to the frequency range of 24 GHz to 100 GHz, which corresponds to wavelengths between 1 mm and 10 mm.

The following is the wavelength chart showing 2.4 GHz versus 5GHz at 1 cycle:

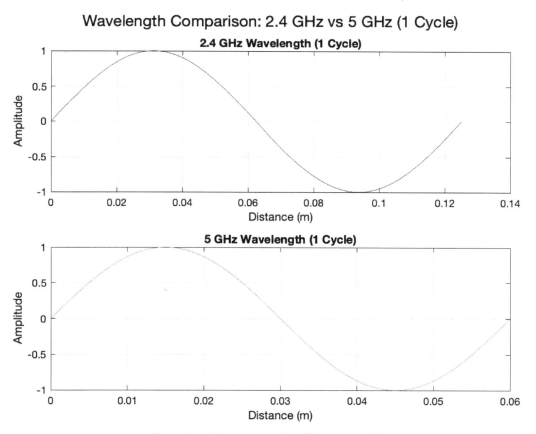

Figure 4.4 – Wavelength of 2.4 GHz versus 5 GHz

Frequency allocation

Typically, the frequency that's managed by a country's regulatory authority is allocated for various purposes, which include both military and civilian uses.

Military frequency:

- Frequencies allocated for military use are often designated for defense and national security purposes

- These may include military communication, navigation, radar, and other specialized military applications

- The specific frequencies that are allocated for military use are usually not publicly disclosed for security reasons

Civilian frequency:

- Civilian frequency refers to the frequencies allocated for non-military use. This broad category includes a variety of commercial, private, and public uses.

- Examples include commercial broadcasting (radio and TV), cellular networks (such as 4G/LTE and 5G), Wi-Fi, satellite communication, emergency services, aviation communication, and more.

- The allocation of these frequencies is typically managed by a national regulatory body (such as the Federal Communications Commission in the United States) and is often made public.

The wireless spectrum that's used by IoT applications falls to the civil frequency.

Spectrum regulation

The electromagnetic wave spectrum in a country encompasses the full range of frequencies, from the lowest to the highest, that are allocated for all types of uses, including civil, military, scientific, and industrial applications. It includes all frequencies that are legally approved and regulated by national or regional authorities. For instance, The **International Telecommunication Union (ITU)** plays a crucial role in coordinating global spectrum allocation, which involves setting guidelines for assigning specific frequency bands for various terrestrial or space radio communication services to prevent international interference. In the United States, the **Federal Communications Commission (FCC)** regulates radio, television, wire, satellite, and cable communications. Similarly, the **European Telecommunications Standards Institute (ETSI)** is responsible for standardizing information and communications technologies in European countries.

Within the civilian usage of the spectrum, there are typically two categories: **licensed spectrum** and **unlicensed spectrum**. Let's take a closer look:

- The **licensed spectrum** requires entities to obtain a license from the country or regional authority for usage. This spectrum is often designated for public and nationwide services such as cellular networks such as 4G/LTE and 5G, as well as broadcast radio and television, where reliable and interference-free communication is essential. Any device transmitting signal within the licensed spectrum must obtain the license from a country regulation authority, such as the FCC in the US.

- The **unlicensed spectrum** refers to radio frequencies that do not require a government license to use, but devices operating in these bands must still follow certain technical standards to minimize interference and make efficient use of the spectrum. The ISM band is a specific part of the unlicensed spectrum that was originally reserved for non-telecommunications purposes such as industrial, scientific, and medical uses. The ISM bands are now widely utilized by various wireless technologies, including Wi-Fi (at 2.4 GHz, 5 GHz, and 6 GHz), Bluetooth, ZigBee, and Thread (all at 2.4 GHz), as well as LoRaWAN and Wireless Hart (typically at 433 MHz, 868 MHz, and 915 MHz).

Spectrum bandwidth

The spectrum bandwidth refers to the width, or range, of frequencies that a communication channel or signal occupies in the electromagnetic spectrum. It is a fundamental concept in telecommunications and wireless communications, indicating the capacity of a channel to carry information.

For instance, Wi-Fi that operates at the 2.4 GHz frequency covers a range of frequencies from 2.4 GHz to 2.4835 GHz. This range is also known as the total channel bandwidth and is approximately 83.5 MHz. Wi-Fi operating at the 5 GHz band covers a range from 5.150 GHz to 5.850 GHz, offering more than 600 MHz of spectrum bandwidth. Please note that the exact frequency boundaries can vary by region due to different regulatory environments.

Channel bandwidth

The channel bandwidth is the frequency range allocated to each communication channel, measured in Hz. It represents the difference between the highest and lowest frequencies within a channel. For instance, according to the **Institute of Electrical and Electronics Engineers** (IEEE) 802.11 b/g/n standards, Wi-Fi operating at 2.4 GHz band ranges from 2.4 GHz to 2.4835 GHz, which offers a total bandwidth of approximately 83.5 MHz. Within this spectrum, the standard defines 11 overlapping channels (in most countries), each with a channel bandwidth of 20 MHz.

dB, dBm, and dBi

Decibel (dB), **decibel-milliwatts (dBm)**, and **decibel-isotropic (dBi)** are essential power units in wireless communication, each serving a specific purpose. dB is used for relative power levels, dBm for absolute power levels, and dBi for antenna gain. A clear understanding of these terms is fundamental for professionals working with wireless systems and networks:

- **dB**:
  - **Definition**: A relative unit of measurement expressing the ratio between two values. It is a logarithmic unit that's used to describe a change in value (such as power or intensity).
  - **Usage**: dB is used to represent gain, loss, and relative levels in signals. It is dimensionless since it describes a ratio.
  - **Example**: Saying an amplifier provides a gain of 10 dB means it increases the power level of the input signal by a factor corresponding to 10 dB.

- **dBm:**

  - **Definition**: An absolute unit of power in decibels, referenced to 1 **milliwatt (mW)**.

  - **Usage**: dBm is used to define a power level on an absolute scale. It is widely used in wireless communications to express transmitter output power, receiver sensitivity, and more.

  - **Example**: A power level of 0 dBm corresponds to 1 mW. A signal with a power level of -10 dBm is less powerful than one with 0 dBm.

- **dBi:**

  - **Definition**: A measure of antenna gain. It compares the power radiated by an antenna to the power radiated by a hypothetical isotropic antenna, which evenly distributes energy in all directions.

  - **Usage**: dBi is used to indicate how effectively an antenna can direct or focus energy in a particular direction compared to the isotropic antenna.

  - **Example**: An antenna with a gain of 3 dBi means it can focus the signal power 3 dB higher than an isotropic antenna in a particular direction.

Let's look at the key differences:

- **dB versus dBm**: dB is a relative measure (ratio) without reference to an absolute scale, whereas dBm is an absolute measure referenced to a specific power level (1 mW; when you say 0 dBm, it means the power level is equal to 1 mW).

- **dBm versus dBi**: dBm measures power levels, while dBi measures antenna gain. dBm is about the signal's strength, and dBi is about how effectively an antenna can transmit or receive signals in a specific direction.

Transmission power

Transmission power, often referred to as transmitter power or TX power, is a key concept in wireless communication. It refers to the amount of electrical power used by a radio transmitter when sending out a signal. Transmission power is typically measured in **watts (W)** or more commonly in mW, and sometimes it is represented in decibels relative to a milliwatt (dBm). For instance, a transmission power of 1 mW is equivalent to 0 dBm.

There are country or regional regulatory limits on how much power can be used for transmission in different wireless bands and applications. These limits are set by organizations such as the FCC in the United States and the ETSI in Europe to avoid interference with other wireless services and to minimize health risks associated with electromagnetic exposure.

Effective Radiated Power (ERP) and **Effective Isotropic Radiated Power** (EIRP) are important concepts in the field of wireless communication, particularly when discussing the power output or strength of a transmitter (TX power). Both terms describe the power level of a radio signal in terms of its effectiveness compared to a reference power. Let's take a closer look:

- **ERP**:

  - **Definition**: ERP refers to the amount of power that a theoretical half-wave dipole antenna would need to emit to produce the same signal strength as the actual transmitting antenna in the direction of its highest gain

  - **Calculation**: ERP is calculated based on the transmitter's output power, the transmission line losses, and the gain of the antenna compared to a half-wave dipole antenna

  - ERP (dB) = TX Power (dBm) + Antenna Gain (dBi) − Line Losses (dB)

  - **Usage**: ERP is commonly used in broadcasting and for VHF and UHF television antennas

- **EIRP**:

  - **Definition**: EIRP is the equivalent power that an isotropic antenna (which radiates power uniformly in all directions) would emit to produce the same signal strength as the actual antenna in its strongest direction.

  - **Calculation**: EIRP is like ERP but uses an isotropic antenna as the reference. It considers the transmitter power, antenna gain relative to an isotropic radiator (dBi), and line losses.

  - EIRP (dB) = TX Power (dBm) + Antenna Gain (dBi) − Line Losses (dB).

  - **Usage**: EIRP is widely used in the context of satellite communication and mobile cellular networks.

Signal strength and quality

Signal strength and quality are critical factors in wireless technology. Understanding how signal strength is measured and its impact on wireless communication reliability and performance is essential, especially for beginners.

Signal strength refers to the power level of a wireless signal that's received by a device from a source, such as a Wi-Fi access point. It is a direct measure of the power level of the received signal, without any specific reference to a scale or index:

- **Measurement**: This is typically quantified in dBm. Higher dBm values indicate a stronger signal.

- **Impact on network performance**: A stronger signal usually translates to improved network performance, evidenced by higher data rates and more stable connections. Weak signals can lead to reduced data rates, connection instability, and a limited range.

Signal quality, distinct from signal strength, focuses on the clarity and integrity of the wireless signal. It encompasses two critical factors, the **Received Signal Strength Indicator (RSSI)** and the **Signal-to-Noise Ratio (SNR)**, both of which are vital for the accurate reception and interpretation of data:

- **RSSI**: RSSI is an estimate of the signal strength that a wireless device receives from an access point. Though expressed in dBm, it is a relative index and not standardized, meaning its values can differ between devices. A higher RSSI value, closer to 0 dBm, typically indicates a stronger signal. For example, -50 dBm is considered a better signal strength than -100 dBm.

- **SNR**: SNR is a crucial metric for assessing signal quality in wireless communications that measures the difference between the signal strength (S) and the level of background noise (N), typically expressed in dB. A higher SNR suggests a clearer signal concerning the noise, leading to better wireless performance. For instance, an SNR value of 25 dB or higher is usually indicative of a robust and reliable Wi-Fi connection. However, external interferences, such as from microwaves or other Wi-Fi networks, can negatively impact signal quality.

The basic formula for SNR in terms of power is $SNR_{dB} = S/N$, where S is the signal power and N is the noise power. Both should be in the same units (for example, W or mW). While SNR is commonly expressed in dB, there are instances where you may need it as a plain ratio. In these cases, you can convert SNR_{db} into SNR_{ratio} by using the $SNR_{ratio} = 10^{\frac{SNR_{dB}}{10}}$ formula.

Shannon's Law and theoretical channel capacity

Shannon's Law, formulated by Claude Shannon, is a fundamental theorem in information theory that defines the maximum data rate (or channel capacity) that can be achieved over a communication channel under given conditions without error. It is crucial for understanding the limits of what can be transmitted over a channel, especially in wireless communication.

Shannon's Law is expressed in the following formula to calculate the maximum channel capacity:

$$C = B \times \log 2 \left(1 + SNR_{ratio}\right)$$

Here, we have the following:

- C is the theoretical channel capacity (maximum error-free data speed in bits per second that can be handled by a communication channel, known as the maximum channel data rate)

- B is the bandwidth of the channel (in hertz)

- S is the average signal power received over the bandwidth calculated in watts (or volts squared)

- N is the average interference power or noise over the bandwidth calculated in watts (or volts squared)

- SNR is the **SNR** of the communication signal to the Gaussian noise interference depicted as the linear power ratio

Shannon's Law implies that the data rate can be increased either by increasing the bandwidth of the channel or by improving the SNR. However, there is a theoretical limit to the data rate that can be achieved even with a remarkably high SNR.

For instance, to assume a bandwidth of 20 MHz and an SNR of 30 dB for a Wi-Fi channel, the maximum channel capacity is calculated as follows:

Convert 30 dB of SNR from dB to ratio: $SNR_{ratio} = 10^{\frac{30}{10}} = 1000$

$C = B \times \log 2 \left(1 + SNR_{ratio}\right) = 20{,}000{,}000 * \log 2 \left(1 + 1000\right) \approx 200 Mbps$

This capacity indicates the theoretical upper limit of data transmission for a 20 MHz Wi-Fi channel, given the SNR at 30 dB.

Modulation

Modulation is a basic process in wireless communication that changes a carrier wave to encode information so that it can be sent through the air. It is important for transmitting data effectively and efficiently over wireless channels. The specific technique that's used depends on factors such as the needs of the application, the distance of the transmission, the data rate, and the conditions of the environment.

The following are the key functions of modulation in the wireless communication process:

- **Digital and analog signals**: Modulation can be applied to both digital and analog signals. In digital modulation, a digital signal (such as binary data) modulates the carrier wave. In analog modulation, an analog signal (such as a voice or music) performs the modulation.

- **Conversion process**: For digital signals, modulation often involves converting these signals into an analog format suitable for transmission over a radio frequency channel. This conversion is essential because electromagnetic waves transmitted through space are inherently analog.

- **Carrier wave**: The carrier wave, typically a high-frequency sinusoidal wave, is modified as per the information signal. This modification can be in terms of amplitude, frequency, phase, or a combination of these aspects.

- **Frequency allocation**: Modulation allows signals to be transmitted over specific frequency channels, allowing for multiple different transmissions to occur simultaneously in different frequency bands, thus efficiently utilizing the electromagnetic spectrum.

- **Environment adaptation**: Different modulation techniques offer varying degrees of resistance to interference and noise, making some more suitable than others for certain environments or transmission distances.

Analog modulation schemes are used to transmit analog information, such as audio or video signals, using a carrier wave of a certain frequency. In analog modulation, some aspect of the carrier wave is varied as per the analog signal's amplitude, frequency, or phase:

- **Amplitude modulation** (**AM**): This modulation scheme changes the strength of the carrier signal based on the message signal. AM is used in AM radio broadcasting, aviation radio, and earlier television broadcasting systems.

- **Frequency modulation** (**FM**): This modulation scheme changes the frequency of the carrier signal in response to the message signal. FM is widely used for FM radio broadcasting and has a higher resistance to noise and interference compared to AM.

- **Phase modulation** (**PM**): This scheme changes the phase of the carrier signal according to the message signal.

Digital modulation schemes are methods that are used in communication systems to transmit digital information using a carrier wave. Unlike analog modulation, which transmits analog signals, digital modulation encodes digital data into the carrier wave. Some examples of popular modulation schemes are provided here:

- **Gaussian Frequency Shift Keying** (**GFSK**) is a modulation scheme that uses a Gaussian filter to shape the frequency spectrum of the transmitted signal. GFSK modulates data by varying the frequency of the carrier wave. The Gaussian filter reduces bandwidth and interference. GFSK is widely used in BLE and low-power radio frequency applications.

- **Phase Shift Keying** (**PSK**) is an important modulation scheme that uses different phase angles to represent symbols. PSK encodes data by changing the phase of the carrier signal. It has various levels, such as **Binary PSK** (**BPSK**), **Quadrature PSK** (**QPSK**), or higher-order PSK (for example, 8-PSK), depending on how many bits are encoded per symbol. The higher the PSK level, the higher the data rate, but also the higher the noise sensitivity and the lower the power efficiency. PSK is commonly used in Wi-Fi, RFID, Bluetooth Classic, and satellite communications.

- **Quadrature Amplitude Modulation** (**QAM**) is an advanced modulation technique that combines amplitude and phase modulation. It enables higher data rate transmission and is widely used in digital communication systems. QAM combines amplitude modulation with phase shift keying, allowing for increased information per symbol. QAM can have diverse levels, such as 16-QAM, 64-QAM, and 256-QAM or higher, depending on how many bits are encoded per symbol. The higher the QAM level, the higher the data rate, but also the higher the noise sensitivity and the lower the power efficiency. QAM is simple to implement and has good spectral efficiency, but it is vulnerable to fading and nonlinear distortion.

- **Orthogonal Frequency-Division Multiplexing (OFDM)** is a frequency multiplexing technique that divides the whole available bandwidth into smaller chunks for efficient data transmission. It splits the total spectrum into many narrow subcarriers, each with a different frequency and not interfering with each other. OFDM can use any modulation scheme, such as QAM or PSK, on each subcarrier, depending on the channel conditions and the data rate required. This means these subcarriers can be modulated independently using different modulation schemes by QAM or PSK. OFDM has several advantages, including high efficiency, resistance to signal fading, and adaptability to different channels. OFDM with QAM is widely used in Wi-Fi 5, 6, and 7, 4G/LTE, and 5G.

Antenna technology

Antennas play a pivotal role in wireless communication systems, serving as the interface between devices and the airwaves. They convert electrical signals into electromagnetic waves for transmission and vice versa for reception. Understanding the fundamentals of antenna operation and design is crucial, especially for those new to wireless technology.

The following are the basic functions of antennas:

- **Converting signals**: Antennas are designed to efficiently convert electrical signals into electromagnetic waves and back. This involves electromagnetic radiation and reception.

- **Resonance**: Most antennas are resonant devices, which means they naturally oscillate at specific frequencies. The size and shape of an antenna typically determine its resonant frequencies.

- **Polarization**: This refers to the orientation of the electric field of the radiated wave. Antennas can be polarized vertically, horizontally, or circularly.

- **Gain**: Antenna gain is a measure of how well the antenna directs or concentrates the signal in a particular direction. Higher gain antennas can transmit and receive signals more effectively at greater distances but usually have narrower coverage areas.

Let's look at different types of antennas and their use cases:

- **Patch antennas**: Patch antennas are compact antennas that are commonly used in IoT end devices, mobile phones, and wireless routers. These antennas are highly regarded for their low profile, which allows them to be easily mounted on flat surfaces. In addition, patch antennas offer excellent performance in terms of signal reception and transmission. Their compact size also makes them ideal for applications where space is limited.

- **Panel antennas**: They are designed to transmit and receive signals in a specific direction, offering increased signal strength and coverage. Panel antennas consist of multiple antenna elements arranged in a panel-like structure, allowing for focused and efficient signal transmission. These antennas are often used in applications such as point-to-point communication, wireless backhaul, and outdoor wireless networks.

- **Omnidirectional antennas**: These antennas are designed to radiate and receive signals in all horizontal directions, providing a wide coverage area. They are commonly used in various applications, including home Wi-Fi routers and base stations for cellular networks. With their ability to transmit and receive signals in all directions, omnidirectional antennas ensure a reliable and consistent signal strength throughout the coverage area, making them an ideal choice for situations where signal coverage needs to be extended in all directions.

- **Dipole antennas**: The simplest form of antenna, they consist of two conductive elements. They are used in various applications, from FM radios to more complex systems.

- **Yagi-Uda antennas**: These are directional antennas that are known for their high gain. They are commonly used in television reception and point-to-point communication links.

- **Dish antennas**: These dish-shaped antennas, also known as parabolic antennas, are highly directional antennas with high gain. Used for satellite communication and radio telescopes.

The following factors will influence which antenna you choose:

- **Frequency of operation**: Different antennas are optimized for different frequency bands

- **Directionality**: Whether the application requires directional or omnidirectional coverage

- **Physical size and form factor**: Constraints based on the device or system the antenna is being integrated into

- **Environment**: Indoor, outdoor, urban, or rural environments can all influence the type of antenna used

Propagation distance

Propagation distance in wireless communication is a measure of how far a signal can travel from its source while maintaining effectiveness and reliability. It is a crucial aspect to consider in the design and deployment of wireless systems as several factors can affect how a signal propagates through space. Let's take a closer look:

- **Transmitter power**: The strength of the signal at its source significantly impacts how far it can travel, but maximum transmitter power is limited by country regulations.

- **Receiver sensitivity**: The ability of the receiver to detect the minimum level of a signal is key to determining the maximum distance. It specifies the characteristics of the wireless chipset from each vendor.

- **Antenna types**: Both the transmitting and receiving antennas' gain, type, and orientation play a role in signal propagation.

- **Frequency of the signal**: Higher frequency signals tend to have shorter propagation distances due to increased attenuation, especially by obstacles.

- **Environmental conditions**: Physical barriers, atmospheric conditions, and interference from other sources can degrade signal strength over distance.

Link budget

In wireless communication systems, we use the link budget to evaluate the distance the signal can travel. It accounts for all the gains and losses in the system, such as transmitter power, antenna gains, path loss, and obstacles.

The link budget has two parts: the **uplink budget** for transmitting data from the user device to the access point, and the **downlink budget** for transmitting data from the access point to the user device. Both budgets limit the actual distance the signal can travel.

The uplink budget (in dBm) can be calculated by the following formula:

$$\text{Uplink Budget} = P_{TX, Up} + G_{TX, Up} - L_{TX, Up} - L_{Pathloss, Up} - L_{RX, Up} + G_{RX, Up}$$

Here, we have the following:

- $P_{TX, Up}$ is the transmitter power of the user device (in dBm)
- $G_{TX, Up}$ is the transmitter antenna gain of the user device (in dBi)
- $L_{TX, Up}$ is the losses in the transmitting system (such as cable loss, connector loss, etc.) of the user device (in dB)
- $L_{Pathloss, Up}$ is the path loss for the uplink (in dB)
- $L_{RX, Up}$ is the losses in the receiving system (such as cable loss, connector loss, and more) of the access point or base station (in dB)
- $G_{RX, Up}$ is the receiver antenna gain of the access point or base station (in dBi)

The downlink budget (in dBm) can be calculated by the following formula:

$$\text{Downlink Budget} = P_{TX, Down} + G_{TX, Down} - L_{TX, Down} - L_{Pathloss, Down} - L_{RX, Down} + G_{RX, Down}$$

Here, we have the following:

- $P_{TX, Down}$ is the transmitter power of the access point or base station (in dBm)
- $G_{TX, Down}$ is the transmitter antenna gain of the access point or base station (in dBi)
- $L_{TX, Down}$ is the losses in the transmitting system (such as cable loss, connector loss, and so on) of the access point or base station (in dB)
- $L_{Pathloss, Down}$ is the path loss for the downlink (in dB)
- $L_{RX, Down}$ is the losses in the receiving system (such as cable loss, connector loss, and so on) of the user device (in dB)
- $G_{RX, Down}$ is the receiver antenna gain of the user device (in dBi)

Comparing the link budget with receiver sensitivity

Incorporating receiver sensitivity into the link budget equation is important when determining if a communication link will work reliably. Receiver sensitivity is the minimum signal strength needed for the receiver to understand the signal. It is important to compare the link budget with the receiver sensitivity. If the link budget is equal to or greater than the receiver sensitivity, the link is viable. In other words, the received signal strength must be equal to or greater than the receiver's sensitivity for reliable communication.

For example, if the downlink budget is -65 dBm and the user device's wireless chipset claims its receiver sensitivity is -90 dBm, the downlink is considered feasible.

By now, you should have a basic understanding of the fundamental aspects of wireless data communications. These are significantly important for the success of IoT applications in actual field deployment.

In the following sections, we will apply this knowledge to popular wireless technologies such as BLE, Wi-Fi, LTE-M, and NB-IoT in the context of 4G/LTE and 5G cellular technologies.

BLE

The concepts of Bluetooth, Bluetooth Classic, and BLE are often misunderstood and confused by beginners. It is important to understand these concepts to fully grasp the capabilities and functionalities of Bluetooth technology.

Bluetooth is a wireless technology that allows devices to communicate and transfer data over short distances. It has become widely popular and is used in various applications, such as connecting smartphones to wireless headsets or transferring files between devices.

Bluetooth Classic, also known as Bluetooth **Basic Rate/Enhanced Data Rate (BR/EDR)**, is the original version of Bluetooth technology. It is commonly used for audio streaming, file transfer, and other applications that require a continuous and steady connection. Bluetooth Classic provides a higher data transfer rate but consumes more power compared to BLE.

BLE is a power-efficient version of Bluetooth technology. It was introduced to address the growing demand for devices with low power consumption, such as fitness trackers, smartwatches, and other IoT devices. BLE is designed for short bursts of data transmission, making it ideal for applications that require intermittent communication and long battery life.

BLE has emerged as the most widely adopted wireless technology in IoT applications, particularly in the domains of smart home and personal wearables. BLE has become a ubiquitous technology that's embedded in billions of devices today, from smartphones and computers to IoT devices and medical equipment. Such significant success stems from its user-friendly connectivity experience for end users, as well as the benefits it provides to IoT innovation developers: low cost, low power consumption, long-range, easy integration, and low effort to quickly deploy in home and individual usage scenarios. With the rapid growth of IoT, BLE has become an indispensable component, enabling massive power-consumption-sensitive IoT devices to communicate and interact with the cloud.

History and current status

The history of Bluetooth technology is quite fascinating and is marked by collaboration, innovation, and the drive to create a universal wireless communication standard. Bluetooth is named after a 10th-century Scandinavian king, Harald "Bluetooth" Gormsson, who is known for uniting Danish tribes into a single kingdom. The name was intended to reflect the technology's ability to unite communication protocols. The development of Bluetooth technology began in 1994 by Ericsson, a Swedish telecommunications company. The aim was to develop a wireless alternative to RS-232 data cables. In 1998, five companies – Ericsson, Nokia, IBM, Toshiba, and Intel – formed the Bluetooth **Special Interest Group (SIG)**, a not-for-profit organization to oversee the development of Bluetooth standards and licensing of Bluetooth technologies.

Bluetooth 1.0 to 3.0 – the age of Bluetooth Classic

The initial Bluetooth specification (version 1.0 and 1.0B) was published in 1999. It utilized *GFSK* as its modulation scheme. GFSK employed a shifting carrier between two frequencies to represent 1s and 0s. This resulted in a data rate of Bluetooth 1.0 limited to 1 Mbps and a range of up to 10 meters. When Bluetooth 2.0 was released in 2004, GFSK was replaced by two newer modulation schemes: π/4 DQPSK and 8DPSK. These schemes used changes in the phase of the waveforms to transmit information instead of frequency modulation. As a result, the data rate of Bluetooth achieved 2 Mbps and 3 Mbps, respectively. Bluetooth 3.0, which came out in 2009, improved the data rate to 24 Mbps by adding support for IEEE 802.11. This release was a significant milestone for Bluetooth as it paved the way for major technological advancements in wireless data communication, providing reliable and high-speed connections for short-range wireless solutions.

However, at that moment, one major issue prevented Bluetooth from being widely used in IoT: power consumption. Bluetooth 1.0 to 3.0, also called **Bluetooth Classic**, were not energy-efficient, causing IoT devices to quickly drain their batteries. This made them impractical for power-consumption-sensitive IoT applications.

Bluetooth 4.0 to 5.3 and beyond – the era of BLE

To meet the growing demand for low-power consumption IoT devices, Bluetooth 4.0 introduced a new type of Bluetooth called BLE in 2010. BLE in Bluetooth 4.0 operates at a lower data rate of 1 Mbps using the GFSK modulation scheme. While this data rate may not be suitable for products such as wireless speakers or video streaming, which require a consistently high data rate, it benefits IoT applications that only need to send small bursts of data at regular intervals or on demand. For example, a fitness wearable device transmits your heart rate to your smartphone through a mobile app every few minutes. By prioritizing low energy usage, BLE enables many IoT applications to be powered by coin-cell batteries for weeks or even months. This makes them practical and efficient.

The release of Bluetooth 5.0 in 2016 marked a significant improvement over the previous BLE standards. Bluetooth 5.0 introduced different PHYs as below and data rates to meet the varying requirements of IoT applications:

- **Standard PHY**: This continued the 1 Mbps data rate from Bluetooth 4.0, balancing speed with power efficiency for standard BLE applications.
- **LE Coded PHY**: This offers a new PHY option for 125 kbps and 500 kbps, specifically designed for long-range communication with potential extended distances of up to 200 meters under optimal conditions. BLE with LE Coded PHY in Bluetooth 5.0 was called **BLE Long Range** (*BLE-LR*).

In 2020, Bluetooth 5.2 was published, introducing a new PHY known as **LE 2M PHY**. This new PHY supports the 2 Mbps data rate. It doubled the data rate compared to the original BLE standard while maintaining low power consumption, which is beneficial for applications requiring both efficiency and speed.

While Bluetooth 5.2 brought enhancements to BLE, including improvements to the LE Coded PHY, which could potentially increase range, the claim of over 400 meters might be theoretical. In practical use, such ranges are often not achievable due to factors such as environmental interference, physical obstructions, and the capabilities of specific Bluetooth hardware.

Bluetooth 5.3, released in 2022, introduced several enhancements aimed at improving the performance, efficiency, and reliability of Bluetooth devices. While it did not introduce major changes like previous versions, it focused on refining existing features.

Standard organization

Bluetooth technology is regulated by global standards created and maintained by the Bluetooth SIG. The Bluetooth SIG is an organization that oversees the development of Bluetooth standards and licenses Bluetooth technologies and trademarks to manufacturers. The Bluetooth SIG also offers certification programs for products to ensure they meet the required standards for quality and compatibility. In addition to complying with Bluetooth SIG standards, Bluetooth devices must also adhere to regional and international regulations regarding wireless communication, such as those set by the FCC in the United States or ETSI in Europe.

In 2000, to create a formal standard recognized by the broader engineering community, the IEEE adopted and standardized the Bluetooth protocol as IEEE 802.15.1. IEEE 802.15.1 provides the protocol and layers definition for Bluetooth, covering aspects from the **Physical Layer** (**PHY**) up to the link layer. It was aimed at ensuring interoperability and defining the technical specifications for Bluetooth wireless technology, particularly for short-range wireless communication.

Ecosystem players

Bluetooth SIG consists of over 35,000 member companies in telecommunications, computing, networking, and consumer electronics. The widespread adoption of these standards has led to a vast ecosystem of Bluetooth-enabled devices. In addition to enormous chipset manufacturers, the Bluetooth ecosystem includes players, contributors, and developers from various domains:

- Consumer electronics
- Healthcare devices
- Smart home
- Smart building
- Location services
- Logistic applications
- Car infotainment

The following table outlines the various specifications:

Specifications	BLE	Bluetooth Classic
Frequency band	2.4GHz ISM band, ranging from 2.402 – 2.480 GHz	2.4GHz ISM band, ranging from 2.402 – 2.480 GHz
Spectrum bandwidth	80 MHz	80 MHz
Channel bandwidth	2 MHz	1 MHz
Channel allocation	40 channels with 2 MHz bandwidth, 3 advertising channels/37 data channels	79 channels with 1 MHz bandwidth
Spectrum spreading	**Frequency-Hopping Spread Spectrum** (**FHSS**)	FHSS
Modulation scheme	GFSK	GFSK, $\pi/4$ DQPSK, 8DPSK
Data rate	LE 2M PHY: 2 Mb/s LE 1M PHY: 1 Mb/s LE Coded PHY (S=2): 500 Kb/s LE Coded PHY (S=8): 125 Kb/s	EDR PHY (8DPSK): 3 Mb/s EDR PHY ($\pi/4$ DQPSK): 2 Mb/s BR PHY (GFSK): 1 Mb/s

Specifications	BLE	Bluetooth Classic
Tx power	≤ 100 mW (+20 dBm)	≤ 100 mW (+20 dBm)
Power consumption	Approx. 0.001 W-0.5 W	Approx. 1W
Rx sensitivity	LE 2M PHY: ≤-70 dBm (BLE 5.2)	≤-70 dBm
	LE 1M PHY: ≤-70 dBm (BLE 5.0)	
	LE Coded PHY (S=2): ≤-75 dBm (BLE 5.0)	
	LE Coded PHY (S=8): ≤-82 dBm (BLE 5.0)	
Propagation distance	10-30 meters (BLE) and 200 meters (BLE-LR)	10-30 meters
Networking topologies	Point-to-point broadcast mesh	Point-to-point
Device pairing	Optional	Required
Communication direction	One-way direction (unidirectional)	Two-way directional (bidirectional)

Table 4.2 – BLE and Classic specifications

For more information, please refer to `https://www.bluetooth.com/learn-about-bluetooth/tech-overview/`.

BLE and Bluetooth Classic are the main wireless technologies that are used in home-based IoT devices. They offer a combination of efficiency and versatility, suitable for various smart home applications. Wi-Fi is also a key technology in residential settings, known for its high data rate capabilities and cost-effective advantages. In the following sections, we will discuss Wi-Fi technology in detail, including its benefits, applications, and its influence on the modern smart home ecosystem.

Wi-Fi

Wi-Fi, which stands for Wireless Fidelity, is a popular and successful wireless broadband technology. It is widely used in homes, schools, and businesses. Wi-Fi has changed how we connect to the internet and has become an important part of our daily lives. It works well with different devices and provides fast internet access, making it the top choice for individuals and organizations. Whether you're browsing the web, streaming videos, or doing business, Wi-Fi is essential. It is reliable, secure, and easy to use, making it an important part of today's digital world. As technology improves, Wi-Fi will continue to get better and remain a leading wireless connectivity solution.

Wi-Fi, when applied in the IoT domain, serves two distinct scenarios. The first scenario is designed to cater to high-bandwidth applications such as video streaming and security surveillance, where a high data rate is crucial. The second scenario addresses the needs of long-range and low-power consumption applications, known as Wi-Fi HaLow. This variant of Wi-Fi is specifically optimized to provide connectivity to devices that require extended coverage while consuming minimal power.

History and current status

In 1999, the debut of 802.11b Wi-Fi marked a significant step, offering up to 11 Mbps data rate. In the early 2000s, Wi-Fi 802.11a and 802.11g increased their data rates up to 54 Mbps. Introduced in 2009, 802.11n Wi-Fi (Wi-Fi 4) significantly improved speeds (up to 600 Mbps) and range, introducing **Multiple Input Multiple Output** (**MIMO**) technology. 802.11ac Wi-Fi (Wi-Fi 5) was released in 2014, and further increased speeds (up to several Gbps) and efficiency, utilizing 5 GHz bands with primary and advanced modulation techniques. Officially launched in 2019, Wi-Fi 6 (802.11ax) offers increased speed (up to 9.6 Gbps), efficiency, and performance, especially in crowded areas. Wi-Fi 6E extends Wi-Fi 6 capabilities into the 6 GHz frequency band, offering more spectrum, higher data rate, and less interference. Wi-Fi 7 (802.11be) is expected to be the next major leap, potentially tripling the speeds offered by Wi-Fi 6.

It has been observed that Wi-Fi 6 is the most popular player in the market at the time of writing. Additionally, in 2020, the FCC in the United States and ETSI have approved the 6 GHz spectrum for Wi-Fi 6E. Looking ahead to the near future, specifically around 2025, Wi-Fi 7 is expected to emerge and thrive in the market.

Wi-Fi 6

Wi-Fi 6, also known as 802.11ax, introduced several key improvements over its predecessor, Wi-Fi 5 (802.11ac). These enhancements are particularly relevant for IoT applications due to their focus on efficiency, capacity, and performance in environments with a high density of connected devices. Here are the major improvements:

- **Increased speed and efficiency**: Wi-Fi 6 offers higher data rates than Wi-Fi 5, potentially up to 9.6 Gbps compared to Wi-Fi 5's 3.5 Gbps. This is achieved through more efficient data encoding, resulting in higher throughput.

- **Orthogonal Frequency Division Multiple Access** (**OFDMA**): Wi-Fi 6 uses OFDMA, which improves wireless network performance by establishing independently modulating subcarriers within frequencies. This approach allows simultaneous transmissions to and from multiple clients, which especially benefits high-density client use cases.

- **Multi-User, Multiple Input, Multiple Output (MU-MIMO)**: Wi-Fi 5 introduced MU-MIMO for downlink only (from AP to client), while Wi-Fi 6 supports MU-MIMO for both downlink (from AP to client) and uplink (from client to AP). In addition, Wi-Fi 6 supports up to eight streams in MU-MIMO, allowing more devices to communicate with the router simultaneously, compared to Wi-Fi 5's support for four streams.

- **Target Wake Time (TWT)**: TWT is a feature that helps devices schedule when to wake up and communicate with the router. This reduces power consumption, which is beneficial for IoT devices and extends their battery life.

- **Basic Service Set (BSS) coloring**: Wi-Fi 6 introduced BBS coloring, which helps in mitigating signal interference. It tags packets with a "color" to differentiate between different networks, thus reducing interference from neighboring networks.

- **1024-Quadrature Amplitude Modulation (QAM)**: Wi-Fi 6 uses 1024-QAM, which allows more data to be packed into each signal compared to 256-QAM in Wi-Fi 5. This increases throughput and efficiency.

- **Improved range and reliability**: Wi-Fi 6 improves signal range and reliability through techniques such as beamforming, which focuses the Wi-Fi signal on specific devices, improving the signal quality and range.

- **Improved security**: Wi-Fi 5 supports WPA2 only, while Wi-Fi 6 supports the latest WPA3, which provides stronger encryption and enhanced protection against certain types of cyberattacks.

These improvements make Wi-Fi 6 more suitable for environments with many connected devices, such as smart homes, smart cities, and industrial IoT setups, where efficiency, capacity, and performance are critical.

Wi-Fi 6E

Wi-Fi 6E represents an enhancement over both Wi-Fi 6 and Wi-Fi 5, primarily through the expansion into the 6 GHz frequency band. Here are the key improvements that Wi-Fi 6E brings to the table compared to Wi-Fi 6 and Wi-Fi 5:

- **Access to the 6 GHz spectrum**: The most significant improvement is the addition of the 6 GHz frequency band. This band is less congested compared to the traditional 2.4 GHz and 5 GHz bands used by Wi-Fi 5 and Wi-Fi 6, leading to less interference and better performance.

- **Increased bandwidth**: The 6 GHz band provides additional non-overlapping channels. This increased bandwidth allows for larger channels (up to 160 MHz), which are ideal for high-bandwidth applications such as high-definition video streaming and virtual reality.

- **Reduced congestion**: Since the 6 GHz band is exclusive to Wi-Fi 6E devices, there is significantly less congestion compared to the crowded 2.4 GHz and 5 GHz bands. This results in more reliable connections and consistent performance.

- **Improved speed and efficiency**: While Wi-Fi 6 already offers higher speeds than Wi-Fi 5, Wi-Fi 6E devices operating in the 6 GHz band can achieve even higher speeds and greater efficiency due to the additional spectrum and wider channels.

- **Better latency**: The lower congestion in the 6 GHz band also contributes to lower latency, which is critical for applications that require real-time response, such as gaming, augmented reality, and certain IoT applications.

- **Enhanced capacity**: Wi-Fi 6E can handle a greater number of devices simultaneously, which is advantageous in densely populated areas or environments with many IoT devices.

It's important to note that to fully utilize the benefits of Wi-Fi 6E, both the router and the client devices need to support the 6 GHz band. Wi-Fi 6E is backward compatible with Wi-Fi 6 and Wi-Fi 5 devices, but these older devices will not be able to operate in the 6 GHz band. Wi-Fi 6E's advancements are particularly beneficial in scenarios where network performance and capacity are critical, such as in smart homes, smart cities, industrial IoT, and dense urban environments.

6 GHz for Wi-Fi 6E spectrum allocations for Wi-Fi by the FCC in the United States and ETSI in Europe have some differences, particularly in the range and extent of frequency bands available for unlicensed Wi-Fi use. Here is a full spectrum comparison crossing 2.4 GHz, 5 GHz, and 6 GHz:

- **FCC (United States)**:

  - **2.4 GHz band**: This band is similar in both regions, typically covering 2.400 to 2.4835 GHz.

  - **5 GHz band**: The FCC has allocated more spectrum in this band compared to ETSI. It typically includes 5.150 to 5.250 GHz (UNII-1), 5.250 to 5.350 GHz (UNII-2), 5.470 to 5.725 GHz (UNII-2 Extended), and 5.725 to 5.850 GHz (UNII-3).

  - **6 GHz band (Wi-Fi 6E)**: The FCC has made a significant allocation in the 6 GHz band, from 5.925 to 7.125 GHz, adding 1,200 MHz of spectrum for unlicensed Wi-Fi use.

- **ETSI (Europe)**:

  - **2.4 GHz band**: This band is like the FCC's allocation.

  - **5 GHz band**: The ETSI allocations in the 5 GHz band are slightly more restricted. It includes 5.150 to 5.350 GHz and 5.470 to 5.725 GHz. The availability of the upper portion of the 5 GHz band (5.725 to 5.850 GHz) varies in different European countries.

  - **6 GHz band (Wi-Fi 6E)**: The ETSI's allocation for the 6 GHz band is different from the FCC's. The EU had proposed to allocate 500 MHz in the lower part of the 6 GHz band (5.945 to 6.425 GHz) for Wi-Fi 6E, which is less than the 1,200 MHz that's allocated by the FCC.

Wi-Fi 7

Wi-Fi 7 (802.11be) is still in development and its final specifications are expected to be ratified by the IEEE in 2024. Based on the proposed features and advancements, Wi-Fi 7 is anticipated to bring significant improvements over Wi-Fi 6, 6E, and 5. Here are some of the key improvements that are expected:

- **Higher data rates**: Wi-Fi 7 is expected to offer significantly higher data rates than Wi-Fi 6/6E, potentially up to 30-40 Gbps, compared to Wi-Fi 6's maximum of 9.6 Gbps. This is achieved through more efficient spectrum use and higher-order modulation (4096-QAM).

- **Wider channel bandwidth**: Wi-Fi 7 may support wider channel bandwidths of up to 320 MHz, compared to the maximum of 160 MHz in Wi-Fi 6/6E. This increase in bandwidth allows for more data to be transmitted at once, further increasing throughput.

- **Improved latency**: Wi-Fi 7 is expected to have even lower latency than Wi-Fi 6/6E, making it more suitable for applications requiring real-time communication, such as gaming, AR, and VR.

- **Multi-Link Operation** (**MLO**): A significant feature of Wi-Fi 7 is expected to be MLO, which allows devices to transmit and receive data across multiple frequency bands (2.4 GHz, 5 GHz, and 6 GHz) simultaneously. This can lead to better reliability, lower latency, and improved load balancing.

- **Enhanced MU-MIMO**: While Wi-Fi 6/6E supports MU-MIMO, Wi-Fi 7 is expected to enhance this feature, potentially supporting more simultaneous streams. This improvement would be particularly beneficial in environments with many devices, such as IoT applications in smart cities or industrial settings.

- **Better energy efficiency**: Energy efficiency is a crucial aspect of IoT devices, and Wi-Fi 7 is expected to introduce improvements in this area. This would be beneficial for battery-powered IoT devices.

It is important to note that while these improvements are significant, the adoption of Wi-Fi 7 will depend on the availability of compatible hardware and the specific needs of users and applications. For IoT, these enhancements could mean more robust, efficient, and high-speed wireless connections, enabling more complex and demanding IoT applications.

Standard organization

The standard organization responsible for Wi-Fi technology is the IEEE. Specifically, the IEEE 802.11 Working Group, part of the IEEE LAN/MAN Standards Committee (IEEE 802), is tasked with developing and maintaining Wi-Fi standards.

The IEEE 802.11 Working Group has developed various Wi-Fi standards over the years, each identified by a unique suffix, such as 802.11a, 802.11b, 802.11g, 802.11n (Wi-Fi 4), 802.11ac (Wi-Fi 5), and 802.11ax (Wi-Fi 6). These standards define the protocols for implementing **wireless local-area networks** (**WLANs**) and are continually updated to improve factors like speed, range, and efficiency.

Ecosystem players

The Wi-Fi Alliance is a non-profit organization that brings together industry leaders to promote and advance Wi-Fi technology. It acts as a central hub for working together, being innovative, and creating standards that ensure Wi-Fi devices work well together. The Wi-Fi Alliance certifies Wi-Fi products to meet industry standards so that people can trust that their Wi-Fi devices are good quality and compatible.

One significant role of the Wi-Fi Alliance is to improve Wi-Fi technology. When new versions of Wi-Fi are developed, the Wi-Fi Alliance works with industry partners to create the technical standards and requirements. This collaboration ensures that new Wi-Fi standards meet the needs of users and businesses, while also considering factors such as performance, security, and power efficiency.

The Wi-Fi Alliance also has a key role in dealing with the challenges and opportunities of IoT. As more devices get connected to the internet, the Wi-Fi Alliance works to make sure that Wi-Fi can handle the growing need for connectivity, while also improving power usage and expanding coverage. This includes the creation of Wi-Fi HaLow, a version of Wi-Fi made specifically for low-power, long-range IoT applications.

In addition, the Wi-Fi Alliance works to make sure that Wi-Fi technology is used in many different industries and areas. They work with companies, service providers, and other groups to teach people about the good things and abilities of Wi-Fi. They want to show how good Wi-Fi is so that more people will use it and it will stay the best way to connect wirelessly.

The following table summarizes the various specifications in this domain:

Specifications	Wi-Fi 5	Wi-Fi 6/6E	Wi-Fi 7
Frequency band	5 GHz	2.4 GHz, 5 GHz, 6 GHz (Wi-Fi 6E)	2.4 GHz, 5 GHz, 6 GHz
Spectrum bandwidth	FCC: 5.150 to 5.250 GHz (UNII-1), 5.250 to 5.350 GHz (UNII-2), 5.470 to 5.725 GHz (UNII-2 Extended), and 5.725 to 5.850 GHz (UNII-3). ETSI: 5.150 to 5.350 GHz and 5.470 to 5.725 GHz	FCC: 2.400 to 2.4835 GHz; 5.150 to 5.250 GHz (UNII-1), 5.250 to 5.350 GHz (UNII-2), 5.470 to 5.725 GHz (UNII-2 Extended), and 5.725 to 5.850 GHz (UNII-3); 5.925 to 7.125 GHz (Wi-Fi 6E) ETSI:5.150 to 5.350 GHz and 5.470 to 5.725 GHz; 5.945 to 6.425 GHz (Wi-Fi 6E)	Same to Wi-Fi 6/6E

Specifications	Wi-Fi 5	Wi-Fi 6/6E	Wi-Fi 7
Channel bandwidth	20 MHz, 40 MHz, and 80 MHz (160 MHz channel is optional)	20 MHz, 40 MHz, 80 MHz, and 160 MHz	20 MHz, 40 MHz, 80 MHz, 160 MHz, and 320 MHz
Channel allocation	5 GHz: 24 channels at 20 MHz bandwidth (may vary based on region)	2.4 GHz: 11 channels at 20 MHz bandwidth (in most countries) 5 GHz: 24 channels at 20 MHz bandwidth (may vary based on region) 6 GHz: 59 channels at 20 MHz bandwidth (may vary based on region)	Same to Wi-Fi 6/6E
Spectrum spreading	DSSS	DSSS	DSSS
Subcarrier spacing	312.5 KHz	78.125 KHz	78.125 KHz
Modulation scheme	OFDM + 256-QAM	OFDMA + 1024-QAM	OFDMA + 4096-QAM
MIMO	MU-MIMO, downlink only	MU-MIMO, downlink and uplink	MU-MIMO, downlink and uplink
Spatial streams	4	8	16
Data rate	Up to 3.5 Gbps	Up to 9.6 Gbps	Up to 30 to 40 Gbps
Tx power	FCC: typically around 1 watt (30 dBm) ETSI: typically around 200 mW (23 dBm)	FCC: typically around 1 watt (30 dBm) at 2.4 GHz and 5 GHz, 6 GHz low power mode up to 500 mW (27 dBm). ETSI: typically around 100 mw (20 dBm) for 2.4 GHz, and typically around 200 mW (23 dBm) for 5 GHz and 6 GHz.	Same to Wi-Fi 6/6E

Table 4.3 – Wi-Fi specifications

For more information, please refer to https://www.wi-fi.org/.

BLE and Wi-Fi have emerged as two highly focused wireless technologies in home settings. They have gained popularity due to their low-cost implementations, free spectrum license, robust ecosystem, and constant evolution to meet user demands and technological advancements. These features make them ideal for a range of smart home applications, from home automation to security systems, and from connected appliances to entertainment systems.

However, the scenario changes when we step outside the home environment. Here, the kings that dominate the IoT field are the licensed spectrum cellular technologies, such as 4G/LTE and the more recent 5G. These technologies offer broad coverage and high data rates, making them suitable for a variety of IoT applications that require wide-area connectivity. This could range from connected vehicles and smart city applications to large-scale industrial automation and remote monitoring systems. In the following section, we will explore how 4G/LTE and 5G work in IoT applications.

4G/LTE and 5G

4G/LTE and 5G are types of cellular networks that use licensed spectrum owned by local telecommunication service providers. They offer users services such as voice and video calling, messaging, and internet browsing on mobile devices. Additionally, these networks are commonly used in various IoT applications, such as high-resolution camera surveillance, high-speed train and fleet connectivity, large-scale asset tracking, and utilities such as water, gas, and electricity metering.

History and current status

The era of cellular network-enabled IoT applications began with the introduction of 3G networks in the early 2000s, which included the **Universal Mobile Telecommunications System** (**UMTS**) and **Code Division Multiple Access** (**CDMA**). 3G networks offered limited data rates in the air up to 384 kbps. The enhanced version over the basic 3G, **High-Speed Packet Access** (**HSPA**), could offer downlink speeds typically up to 14.4 Mbps and uplink speeds up to 5.76 Mbps. 3G advanced versions such as HSPA+ (Evolved HSPA) could push these speeds further, with theoretical downlink speeds up to 42 Mbps and uplink speeds up to 22 Mbps.

The start of 4G/LTE networks in 2010 was a big step forward to improve IoT applications. In addition to providing higher data rates of up to 300 Mbps in the air, the inception of NB-IoT and LTE CAT-M, born from 4G/LTE, has revolutionized IoT applications. It has benefited industries and individuals by enabling the use of interconnected devices with super-low power consumption and wide coverage.

Since 2020, the 5G network has made a grand entrance into the market, aiming to cover three target use cases: **Enhanced Mobile Broadband** (**eMBB**) for higher throughput up to 10 Gbps, **Ultra-Reliable Low Latency Communications** (**URLLC**) for mission-critical industrial applications with up to a 1 ms round-trip delay, and **Massive Machine-Type Communications** (**mMTC**) for higher densities of up to 1 million devices per km^2 space. By 2024, only eMBB is supported in the current 5G network.

The 6G network is currently being researched and conceptualized. It is expected to be developed and deployed in the 2030s. The goals of 6G networks, which are still being defined, are expected to surpass those of 5G. They aim to provide extremely fast data transmission (potentially in the range of **terabits per second (Tbps)**), exceptionally low latency (in the order of microseconds), and widespread connectivity for various applications.

To support battery-powered IoT applications in wide spaces, such as cities and rural areas, NB-IoT and LTE CAT-M are the most popular 4G/LTE technologies that are widely adopted for their advantages of low power consumption, low to medium data rates, and long propagation coverage. According to ABI Research, by 2026, NB-IoT and LTE CAT-M will capture over 60% of the 3.6 billion LPWAN connections.

NB-IoT

NB-IoT (including LTE Cat NB1 and LTE Cat NB2) is designed for IoT devices that require low data amounts, low bandwidth, long battery life, and can operate unattended for extended periods.

NB-IoT operates within a narrow bandwidth of 200 kHz in the LTE spectrum. It can be deployed within existing LTE carriers or in the unused 200 kHz bands previously used by GSM. It uses OFDM modulation for downlink communication and SC-FDMA for uplink communications.

NB-IoT was designed with limited mobility capability for superior power-saving purposes and is suitable for stationary or low-mobility applications such as utility meters. It has developed various use cases, including smart metering (water, gas, and electricity), smart parking, asset tracking, environmental monitoring, smart agriculture, and more. It provides widespread coverage, efficiency, and cost-effectiveness, making it an excellent choice for IoT applications that do not require high throughput or ultra-low latency.

LTE CAT-M

LTE CAT-M (also known as LTE-M or simply Cat-M, including LTE Cat M1 and LTE Cat M2) is similar to NB-IoT but uses a wider bandwidth than NB-IoT at 1.4 MHz. It works with the same LTE infrastructure and bands as regular LTE. LTE CAT-M has faster data rates and supports lower latency in the air, which is better for IoT applications that need to communicate frequently with moderate latency requirements. LTE CAT-M was designed with mobile capability and is good for applications with moving devices such as vehicle tracking and wearables because it supports higher mobility.

Standard organization

The **Third Generation Partnership Project (3GPP)** is a global standard organization that plays a key role in driving the advancements and development of 4G/LTE, 5G, 6G, and beyond cellular-based technologies. Acting as an international organization with members from all over the world, 3GPP includes telecommunications companies, technology providers, and regulatory bodies. They work together to develop and maintain global standards for mobile communication technologies. This collaboration ensures that the latest advancements in mobile technology are accessible and compatible across different networks and devices worldwide, promoting seamless communication and connectivity for users globally.

3GPP uses the approach of roadmap release to define the feature and capability scope of cellular technologies, such as 5G New Radio for eMBB and 4G/LTE Advanced Pro, which was included in Release 15, 5G for uRLLC and Advanced Industrial IoT (IIoT), which was supported in Release 16, Satellite Integration and 5G mMTC, which was scoped in Release 17, and Release 18, which will focus on the introduction of 6G.

Ecosystem players

The ecosystem for 4G/LTE and 5G is the biggest group for industry in the world! It comprises a complex and extensive network of stakeholders, technologies, standards, and applications that collectively define and enable modern cellular communications. The ecosystem includes the following:

- **Standardization bodies and working groups**: 3GPP itself is a collaboration between several telecommunications standards organizations, including GSMA, ETSI, ARIB, TTA, TTC, ATIS, and CCSA.

- **Mobile network operators**: These are the carriers or service providers from global that deploy 4G and 5G networks and offer communication services to end users.

- **Equipment and infrastructure manufacturers**: These are companies that design and manufacture the hardware required for 4G and 5G networks, including base stations, antennas, routers, and other infrastructure components. This also includes manufacturers of devices such as smartphones, tablets, and IoT devices, such as Apple and Google.

- **Chipset and component manufacturers**: These are the companies that produce the semiconductors and components that are essential for 4G and 5G technology, such as modems, processors, and RF components, including Qualcomm, Intel, Samsung, and others.

- **Regulatory and policymakers**: Governmental and regulatory bodies define the legal and policy framework within which 4G and 5G networks operate, including spectrum allocation, privacy regulations, and international agreements.

- **Research institutions and academia**: Universities and research organizations contribute to the 3GPP ecosystem through research and development, often pushing the boundaries of what's possible in cellular technology.

The following table outlines the various specifications:

Specifications	NB-IoT (Release 14, LTE Cat NB2)	LTE CAT-M (Release 13, LTE Cat M1)
Frequency band	800 MHz, 900 MHz, 1.4 GHz, 1.8 GHz, 2.1 GHz	450 MHz, 700 MHz, 800 MHz, 900 MHz, 1.8 GHz, 2.1 GHz
Spectrum bandwidth	200 kHz	1.4 MHz
Channel spacing	180 kHz	1.08 MHz
Modulation scheme	OFDM/SC-OFDM + QPSK/16QAM/64QAM	OFDM + QPSK/16QAM/64QAM
Duplex mode	Half duplex	Half duplex and full duplex
Data rate – downlink	127 kbps	1 Mbps (half duplex), 384 Kbps (full duplex)
Data rate – uplink	159 kbps	1 Mbps (half duplex), 384 Kbps (full duplex)
Latency	1.6 - 10s	50 - 100 ms
Tx power	23 dBm	20 dBm
Range estimate	15 km/s	10 km/s
Mobility	Limited	Yes
Roaming	Limited	Yes
Battery lifetime estimate	Up to 10 years with AA (5000 mAh) battery	Up to 10 years with AA (5000 mAh) battery

Table 4.4 – The NB-IoT and LTE CAT-M specifications

Summary

In this chapter, we learned more about wireless connectivity, which is the invisible but powerful force that allows IoT devices to deliver data. We discussed essential information about wireless data communication, investigated the energy-efficient BLE technology, and talked about the common Wi-Fi networks that connect our devices. Additionally, we explored the abilities and future possibilities of cellular networks such as 4G/LTE and the new 5G, which aim to design the use of IoT through LTE-M and NB-IOT.

In the next chapter, where we'll dive into *the cloud*, we'll shift our focus from how data is transmitted to the platforms where it is stored, processed, and analyzed. The cloud is a digital environment where enormous amounts of data from wireless IoT devices can be used to gain new insights and make decisions. The next chapter will explain how cloud computing platforms work with IoT, offering flexible and scalable resources that enable the development of complex IoT applications. We will discuss how the cloud acts as the central system for IoT, providing the necessary processing power and storage to turn raw data into useful information.

5

The Cloud, IoT's "Superpower Brain"

In the intricate world of IoT, the cloud serves a role similar to the human brain, collecting, processing, and analyzing the vast streams of data generated by countless IoT devices, much like the brain interprets signals from neuron cells.

In this chapter, we will explore the important role played by the cloud in IoT solutions and unveil how the cloud effectively addresses and resolves the challenges that arise with the rapid growth of IoT applications.

In addition, we will discuss how IoT interacts with the cloud, including device management and data ingestion by the cloud. By understanding these integration aspects, you can see how IoT and the cloud work together to provide easy connectivity and data usage in real production scenarios.

We will also talk about several popular communication protocols that facilitate data exchange and interaction between IoT devices and the cloud. Understanding these protocols is crucial for efficient and secure communication in IoT ecosystems.

Lastly, we will go through Amazon Web Services (AWS), which offers powerful services that are widely used for building robust and scalable IoT architectures.

By the end of this chapter, you will have a thorough understanding of the role of the cloud in IoT solutions and the components that contribute to the successful implementation of IoT projects.

The following topics will be covered in this chapter:

- Why is the cloud essential for IoT?
- Integrating IoT with the cloud
- Communication protocols between IoT devices and the cloud
- AWS for IoT

Important tips

During the project practice in *Chapters 13*, *14*, *15*, and *16*, you will program ESP32 to connect to AWS, utilize multiple AWS services to process sensor abnormal events, as well as collect, store, and query sensor data, and then transform the real-time data into a visualization dashboard. The knowledge you'll learn about the cloud here will help you complete those practices.

Why is the cloud essential for IoT?

In *Chapters 3* and *4*, we discussed two important aspects of IoT elements: the devices and the wireless connection. However, we also need to consider another crucial element in the IoT system – the **cloud**.

Undoubtedly, the cloud plays a crucial role in facilitating efficient IoT solutions by offering flexibility, scalability, reliability, and cost-effectiveness in managing large-scale IoT data, thus creating significant business opportunities. It serves as the operation center for all IoT device data. It has been proven that the cloud has effectively tackled numerous challenges that were hindering the growth of the IoT industry, including high initial investments, slow responses to dynamic market growth, inflexibility to new service demands, and limited global customer engagement, all of which are addressed in the following sections.

The pain point of IoT before the cloud

A brutal challenge faced by IoT application providers as their businesses grow is the unpredictable scale of growth of IoT devices and data.

Typically, IoT application providers start their innovation by creating prototypes in their labs. Then, they test them on a small scale at a customer site. If they work well, they can expand and implement the IoT solutions on a larger scale, either in one country or globally. The number of IoT devices can quickly increase from a few dozen to hundreds in a week, or even thousands or tens of thousands in a few months, depending on customer growth. As a result, the data that's collected from these IoT devices can rapidly upsurge from megabytes to gigabytes and terabytes. This rapid growth, although a sign of business success, raises an important question for IoT application providers: How can they build a data platform that can handle such unpredictable traffic increases in an elastic, flexible, and cost-effective way?

Before the cloud became widely adopted, IoT application providers typically purchased servers and physically installed them either at the customer's data center or in a public data center. The important task was to make sure that the servers had enough capacity, including CPU, memory, hard disk drives, and networking bandwidth, to handle not only current but also deal with future traffic demands.

For example, during the initial stage of deploying an IoT application, customers may only need to connect a few dozen sensors. This would require a server capable of handling a data transaction rate of approximately 10 packets per second, a 100 MB hard disk drive, and a 100 Mbps networking bandwidth. However, in the coming weeks, if customers rapidly expand their deployment, the number of sensors could increase to tens of thousands. This up-spike scale requires the server to support a data transaction rate of up to 1,000 packets per second, a 100 GB hard disk drive, and a 1 Gbps networking bandwidth. Unfortunately, the existing servers' hardware capacity and networking bandwidth are unable to handle this surge in traffic, resulting in a decrease in service quality and then a horrible customer experience.

In the past, to tackle such unpredictable traffic up-spike, IoT application providers usually equipped the servers with higher capacity than needed from the early stages. However, this approach led to significant overhead costs for small-scale deployments, making it not a cost-effective solution.

As mentioned earlier, during the early stage of go-to-market, IoT application providers only need to support a small number of customers. They may have enough engineering resources to assist these customers in setting up a data platform, either at the customer's office or a public data center. However, as their business grows beyond a small-scale customer base, the number of customers may rapidly increase, requiring additional resources beyond their current engineering capacity. If the customer base crosses countries, regions, and continents, it becomes quite challenging for their engineering resource to implement contracts promptly.

The impact of the cloud

With the rise of the cloud, the traditional approach of self-hosting data centers has been completely changed. Instead of installing physical server hardware, the IoT application providers can now lease the virtual servers and services from cloud service providers such as AWS. This not only makes the process simpler but also brings many advantages.

One of the key advantages is scalability. With the cloud, IoT application providers can easily scale their infrastructure up or down based on their needs. This means that they can effortlessly handle sudden surges in demand without any hassle. Additionally, this scalability also ensures that resources are allocated optimally, leading to improved performance and efficiency.

For instance, AWS offers a prominent feature called **Auto Scaling**. This feature helps you monitor and adjust the capacity of your application across various AWS services, such as **Elastic Compute Cloud (EC2)**, **Elastic Container Service (ECS)**, **DynamoDB** (a fully managed, serverless, key-value NoSQL database), and **Aurora** (a fully managed relational database engine). You can enhance performance, optimize costs, or strike a balance between them using straightforward scaling plans and recommendations.

The following are some of the benefits of AWS Auto Scaling:

- You can set up scaling quickly and easily using a single, intuitive interface
- You can make smart scaling decisions based on your preference and the service's suggestions
- You can automatically maintain performance and availability, even when workloads are unpredictable or changing
- You can balance capacity across Availability Zones for high availability and resiliency
- You can launch multiple instance types and purchase options (Spot and On-Demand Instances) within a single Auto Scaling group to optimize costs

Furthermore, the cloud enables rapid business scaling and extends the customer reach for IoT application providers, particularly in global expansion efforts. For example, AWS offers a vast global cloud infrastructure that serves as the foundation for extensive and flexible IoT deployments. As of December 2023, AWS spans 105 Availability Zones across 33 geographic regions, with over 600 points of presence and 13 regional edge caches. This extensive network ensures that IoT application providers can confidently deploy their solutions globally, guaranteeing high availability, low latency, and strong fault tolerance, regardless of the geographical distribution of their customer base.

Lastly, cost-effectiveness is the biggest advantage brought by scalability and flexibility. By getting rid of the need for physical hardware installation and upkeep, IoT application providers can lower their initial costs by using the cloud. This smartly shifts their spending from upfront expenses to operational expenses. They only pay for the resources they use, which helps them optimize their spending and allocate their budget more efficiently. This cost-effectiveness lets providers allocate more resources to core application development and innovation, promoting growth and progress in the IoT industry.

Now that we have understood the constraints to IoT growth before the advent of the cloud and the advantages brought by the cloud, let's explore how the cloud works in the game of IoT.

Integrating IoT with the cloud

The integration of IoT and the cloud has two main focuses: device management and data ingestion. This integration allows us to connect to and control multiple devices, as well as collect and process large volumes of data. Through the use of cloud services, customers can analyze and obtain valuable insights from this data, resulting in improved decision-making and increased efficiency. Furthermore, this integration creates new possibilities for innovation and the development of cutting-edge applications that can bring about significant changes in various industries.

Device management

In the realm of IoT combined with the cloud, managing end devices is of paramount importance due to the sheer number and diversity of devices involved. This efficient management ensures operations are smooth, secure, and scalable. It entails authorizing customers' devices associated with their cloud accounts and accessing other data services. The following key functions are involved in end-device management:

- **Registration**:
  - **Device provisioning**: This first step involves connecting the device to the cloud by providing its unique identifiers (such as device ID and type) and metadata (such as location, owner, and organization). This helps establish the device's identity within the IoT network.
  - **Credential management**: Given the importance of security in IoT, this phase focuses on creating and protecting the credentials needed for the device's cloud authentication. This includes access keys, passwords, tokens, and certificates. Managing these credentials carefully is crucial to prevent unauthorized access and potential security breaches.

- **Configuration**:

    - **Operational setup**: Post-registration, devices are configured with default settings such as network configurations, data reporting frequencies, and operational modes, aligning device operation with predefined specifications.

    - **Service policy assignment**: Devices are assigned specific roles and permissions to access the data services or applications. These policies determine what actions, resources, and access levels each device can have, making sure they follow their intended operational scope.

- **Monitoring**:

    - **Health monitoring**: This includes checks on operational status, battery life, and connectivity.

    - **Logging and diagnostics**: Systematic logging aids in issue diagnosis, behavioral understanding, and performance enhancement. Log analysis offers insights into system performance, aiding in issue identification and resolution.

- **Life cycle management**:

    - **Firmware updates (FOTA)**: Regular updates via **firmware over the air** (**FOTA**) are crucial for enhancing functionality, rectifying bugs, and reinforcing security

Data ingestion

To keep the data of IoT devices safe, it needs to be securely sent to the cloud. Data ingestion goes beyond just storing the data. It involves preprocessing, storing, analyzing, and visualizing the data. Let's look at these steps in greater detail:

1. **Data collection**: Retrieve the raw data payload from the protocol messages sent by IoT devices, such as extracting raw data from the payload section of MQTT messages published by IoT devices.

2. **Data preprocessing**: Preprocess raw data through transformation, formalization, filtering, appending, or amalgamating it for streamlined storage. An example of this is converting temperature sensor readings from raw data into a uniform unit or appending the current timestamp to the payload data.

3. **Data storage**: Load the processed data into the databases on the cloud, prepping it for subsequent analysis and presentation.

4. **Data analysis**: Employ techniques such as time series analysis for trend spotting or forecasting, or utilize a machine learning algorithm to correlate data points and spot particular events or situations. You should construct alert systems based on preset thresholds or anomaly detection, subsequently notifying customers through messaging or email.

5. **Data visualization**: Present data in a clear and vivid format in a dashboard or generate regular reports summarizing essential metrics and insights, showcasing the health status of the system.

Now that we understand the functions that the cloud can perform in the overall IoT architecture, let's look at how the cloud talks to IoT devices remotely.

Communication protocols between IoT devices and the cloud

In *Chapter 4*, we discussed the application layer of the OSI model as it applies to wireless connectivity. The application layer, also known as **Layer 7**, interacts with software applications and serves as a data communication component.

Indeed, the communication protocols between IoT devices and the cloud are very important in determining how well IoT solutions work. These protocols are designed to meet the specific needs of IoT ecosystems, such as using low power, sending as little data as possible, and making sure data is transmitted reliably even when the network isn't stable. These protocols are part of the application layer of the OSI model.

Several popular protocols are used in IoT with the cloud today, including **Message Queueing Telemetry Transport (MQTT)**, **User Datagram Protocol (UDP)**, **Constrained Application Protocol (CoAP)**, and **Lightweight Machine-to-Machine (LwM2M)**. Let's learn more about some of these protocols.

MQTT

- **Type**: Application layer protocol.
- **Publish/Subscribe model**: Sensors connect to the MQTT broker and publish their data within topics. MQTT brokers act as proxies to make information available to other parties, such as cloud services, databases, dashboards, and even other devices. These parties act as MQTT clients and fetch the information by subscribing to the published topic.
- **Features**:
  - **Lightweight and efficient**: Provides reliable message delivery, retains messages for future subscribers, and supports last will and testament messages for notifying disconnections.
  - **Quality of service (QoS) levels**: Offers three levels of message delivery assurance (At most once, At least once, and Exactly once).
- **Use cases**: Widely used in IoT for sending data from devices to the cloud and vice versa. It's lightweight and designed for low-bandwidth, high-latency, or unreliable networks.

LwM2M

- **Type**: Application layer protocol.
- Built on top of CoAP (transport layer), it takes advantage of CoAP's lightweight and easy-to-implement nature.

- **Features**:

    - **Device management and service enablement**: Efficient in terms of data transmission and power consumption, supports features such as firmware updates, remote diagnostics, and reporting

    - **Secure**: Supports a variety of security modes, including Datagram Transport Layer Security (DTLS)

- **Use cases**: Specifically designed for remote device management and telemetry within IoT. It's used for managing the life cycle of IoT devices, their configuration, and how they collect telemetry data.

CoAP

- **Type**: Web transfer protocol.

- **Request/response model**: Utilizes UDP to keep the protocol lightweight. Servers make resources available with a URL and clients can make requests of the `GET`, `POST`, `PUT`, and `DELETE` types.

- **Features**:

    - **Designed for IoT**: Supports built-in discovery of services and resources, asynchronous message exchanges, and offers low overhead and parsing complexity.

    - **Supports RESTful services**: Allows interaction between IoT devices in a request/response model.

- **Use cases**: Specifically designed for constrained devices and networks (such as IoT). It's used for machine-to-machine (M2M) applications such as smart energy and building automation.

Each protocol has its strengths and is designed to address specific challenges and requirements in IoT communications. The choice of protocol largely depends on the specific needs of the IoT application, including factors such as network reliability, power availability, data transmission frequency, and payload size. Understanding these protocols and their characteristics is crucial for designing robust, efficient, and scalable IoT solutions.

Understanding common communication protocols between IoT devices and the cloud is crucial for selecting the appropriate option based on your specific use cases. Next, we'll focus on a market leader, AWS, and explore its range of popular cloud services for IoT applications. Please note that some of these services will be utilized in the practice project in later chapters of this book.

AWS for IoT

AWS is widely recognized as the leading cloud platform in the market. It offers a comprehensive suite of advanced services that deliver flexibility and scalability for cloud computing. The range of AWS services related to IoT is diverse and encompasses various functionalities. Some of these services are specifically tailored for IoT applications, while others are commonly used services that also play a critical role in processing IoT data.

In this section, we'll cover some popular AWS cloud services that are widely used in IoT applications. These services, when used together, create a complete system for managing IoT applications, including device management, data collection, storage, analysis, and gaining insights from data. Depending on your project requirements, you can use several of these services to build a complete IoT solution.

AWS IoT Core

AWS IoT Core is a cloud service that helps devices connect and interact with cloud applications and other devices securely and easily. It can handle a large number of devices and messages and securely process and route them to different endpoints.

The following are the key features and benefits:

- **Device connectivity and management**: AWS IoT Core ensures reliable and secure device connection to the cloud, no matter where these devices are located. It supports lightweight communication protocols such as MQTT and HTTP, which are ideal for IoT devices with limited resources.

- **Secure device communication**: It provides authentication and encryption for all device connections, guaranteeing that data is only exchanged between devices and AWS IoT Core when their identity is verified.

- **Message broker**: AWS IoT Core's message broker enables communication between IoT devices and AWS services. It can handle a massive number of messages and route them to the necessary endpoints.

- **Rules engine**: The rules engine allows you to process and transform messages effectively. You can use SQL-based rules to query message content and use the results to route the message to other AWS services. For example, you can trigger a Lambda function, send a push notification, or store data in a DynamoDB table based on incoming data.

- **Device shadows**: AWS IoT Core provides device shadows, virtual versions of devices in the cloud. This feature allows you to keep track of your devices' current and desired states so that your applications can interact with them even when they're offline.

- **Integration with other AWS services**: It seamlessly integrates with other AWS services, enabling you to build IoT applications that collect, process, analyze, and take action on data generated by connected devices without the need to manage infrastructure.

AWS IoT Core supports the following communication protocols with IoT devices as of December 2023:

- **MQTT.**
- **MQTT over WebSockets Secure (WSS).**
- **Hypertext Transfer Protocol – Secure (HTTPS).**
- **Long Range Wide Area Network (LoRaWAN).** To learn more, visit `https://docs.aws.amazon.com/iot-wireless/latest/developerguide/iot-lorawan.html`.

The following are its use cases in IoT:

- **Home automation**: Control and monitor home devices such as lights, thermostats, and security cameras

- **Industrial IoT**: Connect and manage industrial equipment such as conveyor belts, robotic arms, and sensors

- **Smart agriculture**: Gather data from agricultural sensors such as soil moisture sensors, weather stations, and GPS trackers on farming equipment

- **Healthcare monitoring**: Securely process and analyze data from healthcare devices such as wearable health monitors and medical equipment

To learn more about AWS IoT Core working in IoT devices, please refer to `https://aws.amazon.com/blogs/compute/building-an-aws-iot-core-device-using-aws-serverless-and-an-esp32/`.

AWS IoT Device Management

AWS IoT Device Management, working as part of IoT Core, is a service that makes it easy to onboard, organize, monitor, and remotely manage your IoT devices at scale. It provides essential tools and features to keep your devices healthy, manage permissions and settings, and troubleshoot devices remotely.

Here are its key features and benefits:

- **Device onboarding**: It simplifies the process of connecting IoT devices. You can register devices individually or in bulk, and AWS IoT Device Management ensures they are properly connected and authenticated before sending data.

- **Device organization**: It can manage and categorize devices in groups and make bulk changes. This is especially useful for managing settings or deploying updates to multiple devices at once.

- **Remote device management**: It can securely manage devices remotely. It can push updates, reset, reboot, or make changes across your devices without physical access.

- **Monitoring and troubleshooting**: It can monitor device health and activity, set alerts for non-compliant devices, and use diagnostic features to troubleshoot remotely.

- **Access control**: It can integrate with **AWS Identity and Access Management** (**IAM**) for granular permission control. Ensure devices only perform authorized actions.

- **Audit and compliance**: It can audit device fleets to ensure compliance with security policies, as well as maintain the security and integrity of your IoT ecosystem.

Now, let's look at some of its use cases in IoT:

- **Fleet management and maintenance**: Manage and maintain your fleet of IoT devices efficiently, ensuring they are always operational, up-to-date, and secure

- **Software updates**: Remotely update the software running on your devices, ensuring they have the latest features and security patches

- **Device monitoring**: Monitor device health and activity, and set up alerts to be notified about potential issues or required maintenance

- **Security and compliance**: Ensure your devices are compliant with your security policies and best practices, reducing the risk of security breaches

To learn more about the function blocks of AWS IoT Device Management, please refer to `https://aws.amazon.com/iot-device-management/`.

AWS IoT Device Defender

AWS IoT Device Defender, working as part of IoT Core, is a service that helps secure your IoT devices. It checks your IoT configurations to ensure they follow security best practices and provides tools to monitor device behavior and detect abnormalities.

The following are its key features and benefits:

- **Continuous auditing**: AWS IoT Device Defender checks your IoT configurations to ensure they follow security best practices. It checks policies, device certificates, and device communication with AWS IoT Core to identify potential security vulnerabilities.

- **Real-time monitoring and detection**: It monitors your IoT devices for abnormal behavior that may indicate a security threat. For example, if a device starts sending data more frequently or to an unrecognized endpoint, IoT Device Defender alerts you.

- **Alerts and mitigation**: When an issue is detected, AWS IoT Device Defender sends alerts based on your criteria. It can also integrate with AWS Lambda to trigger automatic actions, such as quarantining a device.

- **Security metrics**: It provides predefined security metrics to evaluate your device's behavior. These metrics include authentication failures, connection drops, and messages sent.

- **Integration with other AWS services**: AWS IoT Device Defender works with other AWS services such as AWS IoT Core, AWS IoT Analytics, and AWS IoT Events to create a secure IoT ecosystem.

Let's look at some of its use cases in IoT:

- **Prevent unauthorized access**: Detect and respond to unauthorized access attempts against your IoT devices

- **Ensure device authenticity**: Continuously check and ensure your devices use the correct certificates and policies

- **Monitor fleet metrics**: Keep track of your fleet's security metrics from a central place, making it easier to manage many IoT devices

- **Automated security responses**: Automate responses to potential security threats, reducing the time and resources needed for IoT security management

Please refer to `https://aws.amazon.com/iot-device-defender/` to learn more about the function blocks of AWS IoT Device Defender.

AWS IoT Analytics

AWS IoT Analytics is a managed service that makes it easy to analyze large amounts of IoT data without the usual cost and complexity. It's designed specifically for IoT data and analytics, providing advanced capabilities to analyze data from IoT devices and make informed business decisions.

Here are its key features and benefits:

- **Data collection**: IoT Analytics automates the process of collecting data from IoT devices. You can set up a channel to collect IoT data securely and at scale.

- **Data processing and enrichment**: After data collection, the service allows you to process, filter, and enrich the data. You can transform the data into a format that's ready for analysis.

- **Data storage**: The service provides a time series data store optimized for IoT scenarios with large volumes of time-series data, like sensor readings.

- **Data analysis**: AWS IoT Analytics lets you run queries using a built-in SQL query engine. You can also perform more complex analytics and machine learning using Jupyter Notebooks or connect your data to BI tools such as AWS QuickSight for visualization.

- **Integration with machine learning**: The service integrates with Amazon SageMaker, allowing you to build, train, and deploy machine learning models on your IoT data. You can also use your own pre-trained models.

- **Security and privacy**: AWS IoT Analytics provides control over data access using AWS IAM policies, ensuring that only authorized services and individuals can access the data.

Here are some of its use cases in IoT:

- **Predictive maintenance**: Analyze data from industrial equipment to predict when machines are likely to fail so that maintenance can be scheduled just in time to prevent breakdowns

- **Asset tracking**: Analyze location data to track assets in real time and understand asset utilization patterns

- **Environmental monitoring**: Collect and analyze data from environmental sensors to monitor air quality, water quality, temperature, humidity, and more, as well as make informed decisions based on environmental conditions

- **User behavior analytics**: Understand how users interact with IoT devices to help inform product improvements, targeted marketing, and personalized user experiences

To learn more about the function blocks of AWS IoT Analytics, please refer to `https://aws.amazon.com/iot-analytics/`.

AWS Lambda

AWS Lambda is a serverless service that allows you to run code without worrying about servers. It automatically handles the computing resources such as memory, CPU, and network. This is great for IoT applications because it can easily scale, adapt, and save costs.

Let's look at its key features and benefits:

- **Event-driven**: Lambda responds to events from various AWS services and other applications. In IoT, Lambda can process data from devices in real time.

- **Scalability**: Lambda scales your application by running code for each trigger. It can handle workloads of any size, from a few requests to thousands per second.

- **Cost-effective**: You only pay for the time your code runs and when it's triggered. There are no charges when it's not running. This is beneficial for IoT applications with sporadic or unpredictable device interactions.

- **Integrated with AWS IoT**: Lambda can be triggered by AWS IoT rules, allowing real-time data processing systems to be used.

- **Flexible**: Lambda supports multiple programming languages and can be used for any type of application or backend service.

- **Microservices**: Lambda works well with microservices architecture, allowing you to manage backend services without infrastructure.

Here are its use cases in IoT:

- **Data processing**: Transforming or enriching IoT data before sending it to another service

- **Device provisioning and management**: Automatically managing devices connecting to your network

- **Alerting and notification**: Sending alerts or notifications based on IoT sensor data

Please refer to `https://aws.amazon.com/lambda/` to learn more about the flow of AWS Lambda when used in IoT applications.

AWS Kinesis

AWS Kinesis is designed for data streaming services. It offers services that let you load and analyze streaming data in real time, as well as build custom streaming data applications. It's particularly useful in IoT scenarios where many devices generate a large stream of data that needs to be collected, analyzed, and acted upon quickly.

Here's a breakdown of the different components of AWS Kinesis and their key features and benefits:

- **AWS Kinesis Data Streams**: This service lets you build custom applications that process or analyze streaming data for specific needs. In IoT, you can use Kinesis Data Streams to continuously capture and store large amounts of data from various sources, such as sensors, website clickstreams, financial transactions, social media feeds, IT logs, and location-tracking events.

- **AWS Kinesis Data Firehose**: Kinesis Data Firehose is the easiest way to reliably load streaming data into data lakes, data stores, and analytics services. It can capture, transform, and load data into AWS data stores for near real-time analytics using existing **business intelligence** (**BI**) tools and dashboards. IoT applications often use this service to route data from devices to AWS S3, AWS Redshift, or AWS Elasticsearch Service for further analysis.

- **AWS Kinesis Data Analytics**: This service allows you to analyze streaming data with standard SQL without needing to learn new programming languages or processing frameworks. In IoT applications, Kinesis Data Analytics can help run real-time analytics on data gathered from devices, providing immediate insights and enabling quick decision-making.

- **AWS Kinesis Video Streams**: This service simplifies the process of securely streaming video from connected devices to AWS for analytics, ML, and other processing. IoT devices with video feeds, such as security cameras or drones, can use Kinesis Video Streams to stream video to AWS for analysis and storage.

Here are its use cases in IoT:

- **Real-time monitoring and analytics**: IoT devices generate a lot of data. Kinesis can process and analyze this data in real time, allowing for immediate insights and actions. For example, it can be used to monitor equipment health in industrial settings and send real-time alerts if anomalies are detected.

- **Time series analytics**: For IoT applications that require time series data analysis (such as sensor data analysis), Kinesis can ingest this high-volume data for real-time analytics and subsequent decision-making.

- **Video data analysis from IoT devices**: With Kinesis Video Streams, you can stream video from connected devices to the cloud for analytics, machine learning, playback, and other processing.

- **Data routing**: Kinesis Data Firehose can efficiently move data to various AWS services, such as S3, Redshift, or Elasticsearch, where you can use it for further analysis and storage or to build interactive dashboards.

You can learn more about the flow of AWS Kinesis, when used in IoT applications, at `https://aws.amazon.com/kinesis/`.

AWS DynamoDB

AWS DynamoDB is a highly efficient and scalable NoSQL database service specifically designed for IoT applications.

Let's take a look at its key features and benefits:

- **High performance**: DynamoDB can handle over 10 trillion requests per day and support peaks of more than 20 million requests per second. This makes it perfect for handling a large volume of requests from numerous IoT devices.

- **Flexible data storage**: DynamoDB supports key-value and document data structures. This is ideal for storing diverse types of data from IoT devices, such as sensor readings, user settings, and device states.

- **Fast response time**: DynamoDB provides consistently low latency in the single-digit millisecond range, ensuring real-time access to data for timely decision-making in IoT applications.

- **Secure data storage**: DynamoDB encrypts data at rest by default and offers fine-grained access control using AWS IAM, ensuring authorized access to your IoT data.

- **Fully managed and hassle-free**: DynamoDB is a fully managed service that takes care of hardware provisioning, setup, configuration, replication, software patching, and cluster scaling. This allows you to focus on your IoT application development without worrying about database management.

- **Global data replication**: DynamoDB supports multi-region, fully replicated tables, enabling global access to data with minimal latency in IoT applications.

- **Streams and triggers**: DynamoDB Streams captures item-level modifications in a time-ordered sequence and trigger AWS Lambda functions. This feature can be used to send alerts or perform actions in response to changes in device data.

- **Seamless integration with AWS IoT**: DynamoDB seamlessly integrates with other AWS services such as AWS IoT Core, AWS Lambda, and AWS IoT Analytics, providing a comprehensive IoT solution.

In terms of its use cases, In IoT projects, DynamoDB is commonly used for storing and retrieving device telemetry data, device states, user settings, and managing dynamic content. Its scalability, performance, and integration capabilities make it an excellent choice as a backend for IoT applications.

You can go to `https://aws.amazon.com/blogs/compute/implementing-a-serverless-aws-iot-backend-with-aws-lambda-and-amazon-dynamodb/` to learn more about the flow of AWS DynamoDB when it's used in IoT applications.

AWS QuickSight

AWS QuickSight is a cloud-based BI service that provides easy-to-understand insights, visualizations, and dashboards from your data sources. It's designed to be scalable, serverless, and embeddable, with built-in machine learning capabilities.

The following are its key features and benefits:

- **Interactive dashboards**: It can create interactive dashboards that can be accessed from any device. These dashboards automatically update with new data, which is ideal for real-time monitoring of IoT devices and systems.

- **Easy integration**: QuickSight integrates with various data sources, including popular cloud services such as AWS S3, AWS RDS, AWS Redshift, and AWS Athena, as well as third-party databases. You can easily bring in data from IoT devices stored in these services and analyze it using QuickSight.

- **Machine learning insights**: QuickSight provides machine-learning-powered insights to help uncover hidden trends and outliers, forecast future results, and facilitate informed decisions. For IoT, this means predicting device maintenance needs, optimizing operations, or understanding usage patterns.

- **Serverless and scalable**: As a fully managed service, QuickSight requires no server provisioning or management. It effortlessly scales to support tens of thousands of users, making it ideal for handling data generated by potentially millions of IoT devices.

- **Pay-per-session pricing**: QuickSight offers a unique pay-per-session pricing model, which can be cost-effective for organizations with many users who access dashboards infrequently.

- **Embeddability**: You can embed QuickSight dashboards into apps and websites, providing integrated analytics within IoT applications or platforms.

- **Security**: QuickSight offers robust security features, including row-level security, column-level security, and multi-factor authentication, ensuring secure access and analysis of data from IoT devices and other sources.

In terms of its use cases in IoT, QuickSight simplifies the process of visualizing and interpreting data from various devices. It helps stakeholders understand patterns, monitor systems, and make data-driven decisions. For example, you can create dashboards to monitor environmental conditions reported by IoT sensors across different locations, visualize equipment performance in a factory, or track the usage patterns of smart home devices.

To learn more about the flow of AWS QuickSight when it's used in IoT applications, please refer to `https://aws.amazon.com/blogs/big-data/build-a-visualization-and-monitoring-dashboard-for-iot-data-with-amazon-kinesis-analytics-and-amazon-quicksight/`.

Summary

In this chapter, we discussed the indispensable role of the cloud in IoT applications, its essential functions, prevalent communication protocols, and several AWS services designed for IoT. Understanding these concepts is vital when you're planning your IoT innovations. We will practice some of these AWS services in *Chapters 13* through *16*.

So far, we've covered basic IoT elements from a high-level perspective, including its network, architecture, devices, wireless connectivity, and the cloud. In the upcoming chapters, we'll explore how to employ ChatGPT to accelerate our IoT innovation journey. This will include the areas where ChatGPT can be utilized in the development process, the specific prompting framework and skills for our innovation, and some beginner-friendly IoT hands-on projects with prompting examples. Using these skills, you can instruct ChatGPT to accurately draw diagrams about the IoT application logic flow in terms of your design and generate code that's driven on your hardware prototype, as per your intentions.

Part 2:
Utilizing AI in IoT Development

The second part of the book applies AI, especially ChatGPT, to IoT projects, linking theory with hands-on practice. It introduces the integration approach of ChatGPT into IoT projects, discussing AI's role in conceptualizing and executing IoT solutions, and its transformative impact on innovation. It also highlights the interaction principle with ChatGPT, effective prompt framework options for IoT applications, and the example of instructing ChatGPT to generate C++ code snippets on ESP32 in the context of IoT projects. Practical advice on starting IoT projects is also provided, emphasizing a gradual approach to avoid common pitfalls and ensure sustainable development. Ten detailed project examples demonstrate how ChatGPT can streamline the development process, from sensor integration to data handling. Finally, this part explores using AI tools to create flow diagrams, which are crucial for planning and understanding data flow within IoT projects, serving as a blueprint for coding and assembly phases.

This part contains the following chapters:

- *Chapter 6, Applying ChatGPT in the IoT Innovation Journey*
- *Chapter 7, Recommendations to Start Your First IoT Project*
- *Chapter 8, 10 Beginner-Friendly IoT Projects with ChatGPT Prompts*
- *Chapter 9, Using AI Tools to Draw Application Flow Diagrams*

6

Applying ChatGPT in the IoT Innovation Journey

Hello and welcome to this chapter!

Imagine stepping into a world where your creative ideas can turn into real, working gadgets and solutions that make life easier and more fun. This chapter is your exciting step on an amazing journey where we'll explore how to apply the magical power of ChatGPT, the AI-powered chatbot by OpenAI's GPT **large language model (LLM)**, to accelerate your IoT innovation and rapidly bring your coolest ideas to tangible prototypes.

We will begin our journey with a deep dive into how AI is reshaping our lives, work, and society far and wide. We will learn not only about the impact of AI but also about envisioning its potential to revolutionize every sector around us. From the initial spark of an idea to the implementation's final touches, we'll show how ChatGPT can act as both a companion and a guide, boosting creativity, efficiency, and effectiveness throughout the IoT development process.

We will then outline the fundamental principles to follow when engaging with ChatGPT and developing your effective tactics for interacting with it. Here, you will discover the best practices for developing effective ChatGPT prompt frameworks and patterns, specifically designed for IoT applications. Finally, we will look at some valuable skills that you can use to enhance your experience when utilizing ChatGPT for generating C++ code specifically for ESP32 microcontrollers.

In this chapter, we'll cover the following topics:

- Reshaping the future with AI
- Utilizing ChatGPT in IoT development process
- Interacting with ChatGPT properly
- Best practices for beginner IoT projects
- Generating code snippets on ESP32

Important tips

We will apply the ChatGPT prompting techniques introduced in this chapter to an example project in *Chapter 11*. By consistently practicing these skills in real projects, you will gain a full understanding of the interaction principles and enhance your instructional skills for future IoT projects.

Reshaping the future with AI

Think about a world where technology not only understands you but also responds thoughtfully to your every question or need. This isn't just a scene from a sci-fi movie; it's becoming our reality thanks to advancements in multiple AI LLMs, such as OpenAI's GPT, Meta Llama, and Google Gemini. AI is not just another technical buzzword; it's a revolutionary force that's reshaping how we live, work, and interact across the globe.

AI is revolutionizing businesses, services, and industries worldwide. With its advanced technology and capabilities, AI is gaining popularity and making a significant impact in sectors such as customer service and content creation. Its exceptional ability to understand and respond accurately and naturally to human queries is truly impressive. As businesses around the world recognize the potential of AI, its influence continues to grow, bringing us closer to a future where AI-powered solutions are an integral part of our lives.

Here's how AI is making waves in various sectors, simplifying tasks, enhancing experiences, and opening up new possibilities:

- **Education**: AI acts as a digital tutor, helping students with homework, generating quiz questions, and even customizing learning materials to fit individual needs. It's like having a personal learning assistant who's always ready to help you learn in the best way possible.

- **Healthcare**: In healthcare, AI becomes a friendly assistant, answering health questions, guiding people through symptoms, and supporting mental well-being. It's like a 24/7 health companion, always there to offer advice and comfort.

- **Finance**: AI steps into the financial world to help with customer service, offering advice on banking products, and making investment tips feel more accessible. It's like having a financial advisor in your pocket, ready to demystify the complex world of finance for you.

- **Retail**: Shopping becomes a breeze with AI, which can suggest products, answer queries, and even help you find the perfect gift. It's like shopping with a friend who knows exactly what you're looking for.

- **Travel**: Planning your next adventure? AI can offer travel tips, assist with bookings, and personalize dining or accommodation recommendations, making every trip smoother and more enjoyable.

- **Technical support**: AI can simplify tech support, offering solutions to common issues and guiding users through fixes step by step. It's like having a tech expert by your side, anytime you need help.

- **Creative industries**: For the creatives, AI can brainstorm ideas, help write scripts, or even compose music, turning creative blocks into bursts of inspiration.

- **Legal and HR**: AI demystifies legal jargon and streamlines HR processes, making it easier to navigate workplace questions and legal documents.

- **Accessibility**: AI revolutionizes disability access, facilitating interaction with technology through voice commands, text inputs, visual gestures, and other customizable methods.

- **Public services**: Governments and public agencies can use AI to make information more accessible, automate responses to common inquiries, and improve public service delivery.

These use cases illustrate the wide range of applications for AI, demonstrating its ability to streamline processes and create new opportunities for engagement and innovation across various fields.

After examining the transformative impact of AI across various sectors, we will now focus on its specific role in IoT. In the upcoming section, we will explore the stages of the IoT development journey where ChatGPT's coding skills can help. As we navigate the complexities of IoT projects, from analyzing customer pain points to building prototypes, ChatGPT will prove to be an invaluable assistant. This part of our discussion aims to emphasize how ChatGPT can expedite the development process, enhance problem-solving abilities, and foster innovation by providing actionable insights and programming tasks, as well as facilitating a deeper understanding of user needs and system performance.

Utilizing ChatGPT in IoT development process

To navigate the complex world of IoT development processes, you need to understand both what customers want and what the proper solution looks like. ChatGPT, the AI-powered chatbot, is a helpful tool that can support you throughout the entire process, from brainstorming to prototype development.

Here's how ChatGPT can make your IoT development easier:

- **Pain points analysis**: When engaging with customers, it's crucial to be ready for various kinds of feedback and complaints. These complaints can be expressed through different tones and may sometimes hide the real problems. But with ChatGPT's assistance, you can analyze and summarize customer concerns effectively. This helps you identify the most common challenges or *pain points* they experience. By understanding these pain points, you can make informed decisions and take appropriate actions to address them, ultimately enhancing the overall customer experience.

- **Solution brainstorming**: Once you have identified your customers' primary concerns through a thorough pain point analysis, ChatGPT becomes an invaluable assistant during the brainstorming phase of your IoT project. This phase is all about creativity and leveraging ChatGPT's ability to generate a multitude of fresh ideas.

- **Feasibility studies**: With a promising solution from ChatGPT, calibrated by your insights and expertise, the next important step is to assess whether it can work. This is where ChatGPT also plays a crucial role by conducting thorough studies to determine if your IoT project is feasible. Using its extensive knowledge base of the latest technological advancements, ChatGPT can provide you with a detailed analysis of the technological capabilities required to bring your idea to life. It not only evaluates the potential of the current IoT landscape but also predicts future trends that could impact the success of your project.

- **Risk analysis**: Recognizing that no solution is perfect, using ChatGPT for a thorough risk analysis of your IoT project is a crucial step. This AI-powered audit process carefully identifies potential issues in technical, market, and security aspects that could hinder your project's success. Technical risks may involve problems with hardware reliability, software compatibility, or scalability. Market risks could arise from changing user demands, competition, or regulations. Security risks are especially important, including worries about data privacy, possible breaches, and system vulnerabilities.

- **Architecture design**: After the initial studies to determine if the solution is possible, ChatGPT can help you design the end-to-end architecture. This important phase involves making important decisions about which hardware components to use, choosing communication protocols for device connectivity, and using effective data processing techniques for good performance. ChatGPT can provide information about the latest technologies, compatibility considerations, and industry best practices to ensure your solution architecture is not only functional but also scalable and secure. This step-by-step guidance helps establish a strong foundation for your IoT solution, ensuring successful execution and future scalability.

- **Requirement collection**: After completing the architecture design phase, ChatGPT can help define the **minimal viable product** (**MVP**) for your IoT project. It does this by summarizing a concise list of necessary technical and functional requirements. This process ensures that the core components needed for your project's initial success are identified and prioritized. By using ChatGPT, you can extract a clear set of requirements that address both the foundational technological needs (such as specific hardware components, connectivity options, and data processing capabilities) and the key functionalities that provide value to your end users. This MVP approach simplifies the development process, allowing you to quickly bring a functional version of your IoT solution to the market. It meets critical user needs while also setting the stage for future enhancements and iterations. With ChatGPT's help, you can ensure that your project is built on essential features, maximizing efficiency and effectiveness in addressing the identified pain points and objectives.

- **Hardware prototype build**: When you start building the hardware prototype for your IoT project, ChatGPT can assist you with advice and support. This phase is important for turning your idea into a real product, and ChatGPT can help you make choices and solve problems during prototype development. It can give insights based on industry practices and technology advancements. ChatGPT can assist you in choosing the right components, such as MCU, wireless connectivity, sensors, and power supply, that fit your project's needs and budget. It can also provide advice on assembling them and fixing common hardware issues. Whether you're adding sensors, reducing power usage, or ensuring reliable connectivity, ChatGPT can help simplify the process of building the hardware prototype.

- **Software programming assistance**: Once you have successfully assembled and set up your hardware prototype, using ChatGPT for software programming is a crucial next step. This is where your project truly starts to take shape, transforming from a physical assembly into an intelligent and functional IoT system. ChatGPT can help by generating code tailored to your project's specific application logic. Whether you need to program device interactions, process sensor data, manage connectivity, or develop user interface components, ChatGPT can provide code snippets and programming insights that align with your project's requirements.

- **Cloud platform integration**: Finally, an essential aspect of your IoT project involves ensuring that sensor data seamlessly transitions from your hardware devices to a backend cloud platform, such as AWS. This integration is pivotal for harnessing the full scope of cloud computing benefits, including scalable data management and advanced analytics capabilities. To facilitate this crucial step, you can leverage ChatGPT to craft a detailed, step-by-step guide tailored specifically to navigating the intricacies of cloud platform integration.

Navigating the complexities of IoT development, from initial idea generation to integrating with cloud platforms, shows the valuable role of ChatGPT in turning ideas into practical solutions. At every step of the process, such as identifying problems, brainstorming solutions, and conducting feasibility studies, ChatGPT demonstrates its many capabilities in improving efficiency, creativity, and decision-making. However, to fully unlock the potential of this AI tool, it's important to not only have access to ChatGPT but also know how to use it effectively. In the upcoming section, we will explore the details of making the most out of ChatGPT. The aim is to empower you with the knowledge and skills to fully utilize ChatGPT's capabilities, making it a powerful tool for innovation rather than an underutilized resource.

Interacting with ChatGPT properly

Now that we understand the ubiquitous presence of ChatGPT within the IoT development landscape, it's crucial to acknowledge a fundamental truth: the effectiveness of this advanced AI tool hinges not on its capabilities, but on how we interact with it.

The purpose of this section is not to diminish the value of ChatGPT but to provide you with the necessary skills to effectively utilize this tool. By understanding how to create specific prompts, accurately interpret outputs, and integrate insights into your IoT projects, you can transform ChatGPT from a basic digital assistant into a powerful innovation tool. Together, we will explore strategies to enhance your communication with ChatGPT, ensuring that every question serves a purpose and every response helps you achieve your IoT goals. This guide will not only assist you in using ChatGPT but also empower you to master the art of prompting and turn potential into reality.

The following are the fundamental principles when engaging with ChatGPT:

- **Interacting with ChatGPT as a conversation**: Instead of just dumping your questions and waiting for answers, interacting with ChatGPT is a conversation, or a dialog, not a monolog. Unlike using a search engine such as Google, where you enter a query and have to browse a long list of results that may or may not directly and effectively address your question, interacting with ChatGPT requires a more nuanced approach, similar to having a conversation. Imagine that you're meeting someone new, such as a friend, partner, or mentor, and you want to receive meaningful help from them. A clear, precise, and thoughtful instruction, such as well-rounded and considerable opening remarks, will guide the following conversation in a meaningful direction and produce outcomes aligned with your expectations.

- **Don't expect a perfect answer from first-round dialog**: It is not surprising that the first response from ChatGPT may not always perfectly match your expectations or needs. It is crucial to understand that ChatGPT is powered by an LLM framework, utilizing a predictive mechanism to generate responses. This mechanism, rooted in the transformer architecture of LLM, aims to forecast the most likely next word or phrase based on the context provided by you. As such, its answers are shaped by patterns and data from its extensive training, rather than intuitive understanding. This means that some extent of deviation from your expectations always exists. To address this, continuous calibration through ongoing conversation is necessary. Therefore, it is essential to be prepared to refine your queries or ask follow-up questions based on the initial answers. By doing so, you can ensure that you receive the most accurate and relevant information from ChatGPT.

- **Be specific with what you're asking**: The more specific and detailed your requests and contexts are, the more accurate and helpful ChatGPT's responses will be. To ensure you receive the most accurate and helpful response, it is important to provide a well-organized prompt framework and pattern that is specific and detailed. Doing so not only gives ChatGPT a clearer understanding of your request but also guides it to tailor its information to fit your specific needs accurately. Therefore, when engaging with ChatGPT, aim to be as descriptive as possible, incorporating relevant details and background information into your query. This practice of providing comprehensive context ensures that the insights you receive are directly applicable and fully informed, maximizing the utility of your interaction with the model.

- **Mitigate the limits of ChatGPT's knowledge base**: Although ChatGPT is a powerful tool, it does have limitations. The training data for ChatGPT 4 only includes information up until December 2023. Therefore, any events, developments, or updates that have occurred after December 2023 may not be reflected in its responses. To mitigate the knowledge of ChatGPT, you have the option to upload the latest documentation to it for reference. For example, you can paste the ESP32-C3 or ESP32-C6 datasheet, specification, and pin map to ChatGPT during your conversation with it.

- **Encourage creative and diverse responses**: Don't be surprised if ChatGPT provides answers that are different from your expectations. To get the best out of ChatGPT, you should create an open environment that values and encourages different and creative responses. This will help you discover new and innovative solutions. During brainstorming sessions, it is important to embrace and support ChatGPT's creativity as it can lead to breakthrough ideas and unique approaches. You should foster a mindset of openness and appreciation for diverse perspectives so that you can generate imaginative and varied outputs from ChatGPT.

- **Protect your privacy and security**: When using ChatGPT, it is very important to always prioritize your privacy and security. It is strongly advised that you avoid sharing any confidential or private data during conversations. To ensure the highest level of protection, it is recommended to be cautious and not disclose sensitive personal information during your interactions.

- **Continuous validation and cross-checking**: While ChatGPT can provide helpful information and assistance, it's important to remember that it should not be the only source relied upon for making important decisions. It's recommended to verify any information or advice received from ChatGPT with other sources before taking any significant actions. This is especially critical for important decisions. By double-checking and validating the information, you can ensure its accuracy and enhance the credibility of your decision-making process. Using a reliable validation framework allows for a more informed, secure, and confident approach to applying the insights gained, preventing potential mistakes and inaccuracies.

As we explore the fundamental principles of interacting with ChatGPT, we are now at the point where we can dive deeper into the skills of engaging with this AI. By following best practices for interacting with ChatGPT, being specific in our requests, role-playing strategically, and continuously validating and cross-checking, we can have more meaningful discussions about using ChatGPT in our tasks.

In the next section, we will focus on applying these principles practically to IoT projects. This section will guide you in creating effective prompts that address the unique challenges and opportunities of IoT development. Our goal is to ensure that your interactions with ChatGPT go beyond simple conversations and become stepping stones toward innovation and problem-solving in the IoT field.

Best practices for beginner IoT projects

OpenAI has published a comprehensive guide on ChatGPT prompt engineering that's accessible at `https://platform.openai.com/docs/guides/prompt-engineering`. Those strategies and tactics serve as invaluable guidance for anyone eager to master the art of designing effective prompts for LLMs such as ChatGPT 4, aiming to improve interaction quality and tailor responses to specific needs.

Prompt framework options

Many blogs have compiled different effective prompt frameworks that serve as excellent starting points for beginners. These frameworks are systematically categorized based on their focus: **role-led**, **task-led**, **context-led**, and **goal-led** or **purpose-led**. It is crucial to understand and apply the appropriate prompt framework based on the specific focus to effectively communicate your objectives, particularly when guiding AI or team members in project development, research, or problem-solving scenarios. Let's explore how each of these frameworks can be utilized for specific objectives, taking into consideration their strengths and applications.

Role-led framework

In the context of ChatGPT prompting, the role-led framework involves defining prompts based on specific roles or personas that the AI model is expected to emulate. This approach focuses on tailoring the AI's responses according to the characteristics, expertise, and behaviors typical of the role it assumes.

Here are the key aspects of the role-led framework:

- **AI's role definition**: This involves clearly defining the role that the AI will play in your conversation. For example, in the context of an IoT project, you could instruct ChatGPT to act as an experienced software developer proficient in embedded C++ on RISC-V architecture MCU, or in Python to develop algorithms on AWS Lambda.

- **Your persona creation**: This involves describing your persona to the AI with a more complex identity. This persona can possess unique traits, backgrounds, and expertise that shape how it interprets and responds to prompts. For example, you might describe yourself as a high school student with basic Python programming skills.

The following examples are popular formations that highlight the role at the beginning of a conversation:

- **Role, Task, and Format (RTF)**: This framework is simple and straightforward. It tells AI what role it should play, provides background information, and outlines the expected outcome. It's ideal for projects where the individual's role significantly influences the outcome – for example, a skilled software developer writing educational C++ snippets to mentor a high school student who has experience with Python.

- **Role, Action, Context, and Expectation (RACE)**: This framework is action and context-oriented, making it ideal for tasks where the sequence of steps and the environment are important. It focuses on the actions required, the specific context in which they occur, and the expected outcomes. This makes it exceptionally effective in complex scenarios requiring nuanced interactions. For instance, an AI, acting as a proficient system architect, can design an IoT solution to address a customer's existing pain points and meet their expectations. The framework emphasizes the AI's actions within a given context and aims to clarify the anticipated results of these actions.

- **Role, Input, Situation, and Expectation (RISE):** This framework focuses on input and immediate situations, making it ideal for scenarios that require adaptability and instant responses. It clarifies the role of the AI, the input it will receive, the situation it addresses, and the expected outcomes. This framework is particularly beneficial when the AI must handle sensitive or dynamic situations accurately. For example, AI, acting as an IoT application designer, designs a service flow based on specific event occurrences as input and then generates the corresponding reaction.

- **Role, Objective, Scenario, Solution, and Steps (ROSSS):** This framework provides a detailed approach to structuring prompts, focusing on comprehensive task execution and problem-solving. It is particularly useful in complex environments where precise, methodical responses are needed. It helps in creating prompts that not only guide the AI on what to do but also how to do it, encompassing a full plan from problem understanding to solution implementation. For example, AI, acting as an IoT endpoint stack software developer, can be used to develop a communication protocol using TLS and MQTT stack that can talk to AWS.

Task-led framework

The task-led framework focuses on the specific tasks that the AI needs to perform. It is more about the actions and processes involved in completing a task. Thus, the prompts are designed to be highly directive, specifying exactly what the AI is supposed to accomplish.

The following are popular formations:

- **Task, Action, and Goal (TAG):** This is effective for straightforward projects with a clear goal, such as creating a marketing campaign or developing a new feature.

- **Task, Requirement, Expectation, and Format (TREF):** This is ideal for assignments with specific requirements and expected outcomes, such as writing a report or creating a presentation.

- **Task, Requirement, Action, Context, and Examples (TRACE):** This is great for educational or training scenarios, where providing context and examples can help illustrate complex tasks or concepts.

In essence, while the task-led framework is more about *what* the AI does (focusing on tasks and results), the role-led framework is about *how* the AI does it (focusing on behavior and interaction).

Context-led framework

The context-led framework for AI prompting emphasizes the importance of the environment or situation in which interactions occur. It's designed to adapt AI responses based on the specific circumstances surrounding each interaction, ensuring that the AI's behavior is appropriate and relevant to the given context.

The following are popular formations:

- **Context, Task, and Format (CTF):** This is perfect for projects that are heavily influenced by external factors, such as adapting a product for a new market or modifying a service based on user feedback.

- **Context, Action, Result, and Example (CARE):** This is useful for case studies or retrospective analyses, where understanding the action taken and the result in a specific context is crucial.

- **Scenario, Problem, Action, and Result (SPAR):** This is suited for problem-solving scenarios, especially in project debriefs or when planning corrective actions in project management.

- **Scenario, Complications, Objective, Plan, and Evaluation (SCOP):** This is excellent for projects that are anticipated to encounter significant challenges. It's suitable for planning under uncertainty, such as new product development in a highly competitive market. The evaluation component ensures that there is a built-in review process to assess outcomes against objectives.

- **Situation, Task, Action, and Result (STAR):** Commonly used in behavioral interview questions, this framework is also effective for project debriefs and case studies. It helps in dissecting how specific actions taken in particular situations led to outcomes, making it a great learning tool for teams to understand decision-making impacts.

- **Situation, Action, Goal, and Expectation (SAGE):** This is best suited for strategic planning and goal-oriented actions where the situation dictates a specific set of actions to achieve a goal. It's ideal for setting clear expectations for the outcomes of these actions.

The main goal of the context-led framework is to enhance the AI's sensitivity to nuances, improving its interactions by making them more aligned with the user's current situation and broader environment. This leads to more intuitive, helpful, and empathetic AI systems that seem more aware and capable of dealing with complex human environments.

Goal-led framework

The goal-led framework in AI prompting is centered around defining clear objectives that the AI system is meant to achieve in its interactions with users. This approach focuses on the results or outcomes that are desired from the AI's performance, guiding the AI's behavior and decision-making process toward these goals.

The following are popular formations:

- **Goal, Request, Action, Details, and Examples (GRADE):** This is effective for goal-oriented projects with specific requests, such as launching a product or achieving a sales target, where details and examples can guide execution

- **Purpose, Expectations, Context, Request, and Action (PECRA):** This is best for projects where understanding the broader purpose and context is essential for meeting expectations, such as strategic initiatives or organizational change efforts

Each of these frameworks helps organize and simplify complex information, making it easier to understand and communicate. By selecting the appropriate framework for the situation, leaders and teams can improve the planning, execution, and evaluation of their projects or tasks. Whether you're dealing with strategic planning, operational issues, learning and development, or problem-solving, these frameworks offer a structured approach to achieving clarity and alignment.

Some of the frameworks may overlap with one another to some extent. However, these overlaps are not a flaw but a feature that makes them more useful and applicable in different scenarios. By understanding the main components and the specific focuses of each framework, you can better adapt your approach to match the requirements of your projects or tasks.

Best practice prompt examples

Although we talked about many prompt frameworks, it is strongly suggested that beginners starting IoT projects start with the role-led and task-led frameworks, especially RTF, RACE, TAG, and TREF. Let's take a closer look:

- **RACE**: You're acting as a seasoned product manager driving IoT innovation in smart home technology. Your chart identifies customer pain points and translates them into actionable insights to drive new product and solution launches (*Role*). You are dedicated to crafting a comprehensive product requirement document that details not only the MVP features but also conducts an in-depth analysis of the target market, competition, and user experience (*Action*). Your current customers are complaining about …….. (*Context*). Your document will guide your engineering team in developing innovative solutions that resonate deeply with your customers' needs. You will collaborate closely with both the software and hardware development teams to ensure alignment with technical feasibility and resource availability (*Expectation*). Innovation is always your strong desire; you are encouraged to take time to think out of the box (*give ChatGPT time to think*).

- **RTF**: Let's assume you're an experienced solution architect with a comprehensive background in IoT technologies, specializing in smart home technologies (*Role*). Leveraging your deep understanding of industry best practices, you are going to design a robust, cost-effective, and secure IoT solution guided by the provided product requirement document (*using a reference source*). Your design will be documented in the system requirements document and will cover end-to-end functionality (*Task*). You should also include factors such as system security, data privacy, device management, and service scalability. Your document should be organized into four sections, including a system high-level diagram, end device, wireless connectivity, and cloud platform integration. Each section should provide details about the components, key technologies, protocols, recommendations, and the pros and cons (*Format*). System robustness should always be your top priority. You're encouraged to refer to widely adopted architectures in the market today and propose your design after a comprehensive assessment (*give ChatGPT time to think*).

- **TAG**: You're assuming the role of a seasoned senior hardware development engineer with extensive experience in IoT and smart home technologies. Your current mission is to design an innovative smart smoke detector for residential use, focusing on alerting users through SMS and email when they are out of their homes (*Task*). You will develop a hardware prototype reference guide (*Action*) while adhering to the specifications outlined in the system requirements document. For detailed information on chipset capabilities, consult the ESP32-C6 datasheet at `https://www.espressif.com/sites/default/files/documentation/esp32-c6_datasheet_en.pdf` and the MQ-2 smoke sensor specifications at `https://components101.com/sensors/mq2-gas-sensor` (*using a reference source*). Your guide should not only address the technical specifications but also explore innovative design and functionality that enhance user experience. Your exploration of cutting-edge solutions and attention to detail will be pivotal in achieving a balance between reliability, cost, and user-centric design (*the goal and to give ChatGPT time to think*).

- **TREF**: You're a proficient senior software developer who has been tasked with developing the cloud registration feature of a smart smoke detector (*Task*). To implement this feature, the following steps must be taken to meet the service logic (*requirements and specifying the steps needed to complete a subtask*):

  I.   Preconfigure the ESP32 device in AWS IoT Core.

  II.   Initiate the Wi-Fi connection from the ESP32 device to the home Wi-Fi router.

  III. Securely connect with AWS IoT Core through TLS.

  IV.   Create a JSON payload that includes the device ID, model, firmware version, battery level, and current timestamp in the Pacific Time Zone.

  V.   Send the MQTT Publish Topic with a JSON payload to AWS IoT Core.

  VI.   Change the ESP32 device's LED to always-on green upon successful registration.

  VII. Keep the Wi-Fi connection alive for 1 minute.

  VIII. Drop the Wi-Fi connection but still keep the LED always on green.

  IX.   Restart from *Step II* after four hours.

Please structure your C++ source code so that it includes sections for comments, library imports, variable declarations, function definition, a `setup()` function, a `loop()` function, and a line with `*` inserted to separate each section (*Format*).

In this section, we explored the best practices and advanced tactics that are specifically tailored for beginners starting their journey with ChatGPT in IoT projects. The next section will provide valuable tips and examples on effectively requesting ChatGPT to generate C++ code for the ESP32 platform. These insights aim to enhance your understanding, particularly within your engineering examples from *Chapter 11*, and ensure a more streamlined and successful project development process.

Generating code snippets on ESP32

To optimize the use of ChatGPT for generating code snippets on ESP32, especially for complex projects such as your IoT end-device development for smart home use cases, adopting a structured approach to your prompts can significantly improve the output's quality and relevance.

Here are some useful tips that can guide you in crafting effective prompts for generating code:

- **Declare your integrated development environment (IDE) and platform**: Numerous IDEs are available in the market today. It is essential to carefully consider and select the most suitable IDE based on your background and capability. By choosing the proper IDE, you can ensure compatibility with your programming language of choice and harness the full potential of the development environment. Each IDE offers a unique set of features and tools that can greatly enhance your development workflow.

  Example: `I am using Microsoft Visual Studio Code with the PlatformIO extension, espressif32 platform, and Arduino framework.`

- **Specify ESP32 and sensor types**: Before instructing ChatGPT to generate code, it is important to specify the type of device. This step ensures that the code that's generated is tailored to the specific characteristics and capabilities of the device being targeted. By providing this information upfront, developers can optimize the code for efficiency, compatibility, and overall performance. Additionally, specifying the device type allows for better error handling and debugging as any device-specific issues can be identified and addressed during the development process. Taking the time to accurately define the device type at the outset is crucial for producing high-quality and reliable code that meets the intended requirements and objectives.

  Example: `I am using ESP32-C3 as the MCU and MQ2 as the Smoke Sensor in my project.`

- **Define the pin connection**: To ensure accurate pin connections, it is essential to understand the pinout diagrams of components from their datasheet specifications. These diagrams provide detailed information about the pin configurations, enabling you to establish the correct connections. In addition to that, another useful resource is ChatGPT, which can provide recommendations and assistance regarding pin connections.

  Examples:

  - `I am using GPIO2 of ESP32-C3 to connect to the MQ2 Digital Out Pin, and GPIO 4 of ESP32-C3 to connect to the MQ2 Analog Out Pin.`

  - `Please refer to the pin layout map of ESP32-C3 and MQ2 and recommend the pin connection.`

- **Request code structure and comments**: To ensure the code gets structured in a specific manner and includes comments, it is important to clearly request these requirements. By explicitly asking for the code to follow a particular structure, it will enhance readability and maintainability. Additionally, including comments throughout the code will provide valuable explanations and insights for future developers who may need to work on the codebase.

  Examples:

  - ```
    I am an entry-level software programmer working with Python
    and with very little knowledge of C++. Please generate code
    that is easy to understand for someone with my background.
    ```

  - ```
    I'd like the format that each code snippet always starts with
    a brief explanation of its purpose and includes line-by-line
    comments to clarify the logic.
    ```

- **Don't produce pseudocode**: Sometimes, when using ChatGPT, you may come across situations where pseudocode is provided as an example. This pseudocode serves as a representation of how the code could be structured or implemented. However, if you require the actual code with a specific syntax and functionality, you can explicitly request it from ChatGPT.

  Example: `Please generate the actual code for my asks, not pseudo examples.`

- **Gain advice on any security risks in your code**: You may ask ChatGPT if there are any security risks in your code.

  Example: `Please advise me if there is a security concern in my code when including my AWS account access certificates.`

- **Ask for testing and validation strategies**: Request specific strategies for testing and validating each subtask.

  Example: `At the bottom of your code snippets, please include a section about how to validate these code snippets.`

- **Confirm then calibrate**: Sometimes, ChatGPT generates code that is mostly aligned with your expectations but may miss something. If you want to improve the output, you should first confirm the current answer and then provide additional prompts to help calibrate the answer.

  Examples:

  - ```
    I like your answers, it is almost what I expected, let us make
    some improvements based on the current output.
    ```

  - ```
    I like your answers and output format, it is exactly what I
    want, from now on, please maintain this format for future code
    generation cases, ensuring clarity and detailed comments with
    each section clearly explained!
    ```

- **Version control**: You may consider maintaining versions of the generated code using a version control system such as Git. This can help you track changes and facilitate collaboration.

Let's go through a code snippet that generates examples:

Hi, ChatGPT, please act as a proficient senior software developer to develop the cloud registration feature of smart smoke detector. I am a high school student with basic knowledge of Python but no C++ experience. Please ensure that your code is easy to understand for someone with my background. Please try to add comments to each line of your code, do not produce pseudo code.

You shall use Microsoft Visual Studio Code IDE with the PlatformIO extension, espressif32 platform, and Arduino framework. Temporarily hardcode Wi-Fi access credentials for development purposes, but plan for a secure credential management method in production.

The expected feature logic is strictly following these steps:

- Step 1: Pre-configure the ESP32 device in AWS IoT Core

- Step 2: Initiate the Wi-Fi connection from the ESP32 device to the home Wi-Fi router

- Step 3: Securely connect with AWS IoT Core service through TLS

- Step 4: Create a JSON payload including device ID, model, and firmware version, battery level and current timestamp at Pacific Time Zone

- Step 5: Send the MQTT Publish Topic with JSON payload to AWS IoT Core

- Step 6: Change the ESP32 device LED to always-on green upon successful registration

- Step 7: Keep Wi-Fi connection alive for 1 minute

- Step 8: Drop the Wi-Fi connection but keep the LED always-on green

- Step 9: Restart from Step 2 after 4 hours

Please structure your C++ source code including sections for entire code explanation, libraries import, variables declaration, function definition, setup() function, loop() function, and insert a line with "*" between each section.

- Code explanation Section: Begin with comprehensive comments describing the creation date, author, the purpose and structure of the code, security risks, testing and validation approach.

- Libraries Import Section: Import necessary libraries, commenting on the purpose of each library

- Variables Declare Section: Declare the variables used in this code, commenting on the definition of each.

- Functions Section: Define functions with clear comments explaining their purpose and usage.

- Setup Section: Detail the setup() function, including initial configurations.

- Loop Section: Outline the loop() part, describing its continuous operations.

ChatGPT's output may look like what's shown at `https://github.com/PacktPublishing/ Accelerating-IoT-Development-with-ChatGPT/blob/main/Chapter_6/main. cpp`. You may refer to it when you have your own code.

As highlighted earlier, when using ChatGPT to generate code for your IoT project, you must select a prompting framework that's suitable for your situation. Since ChatGPT may not always provide completely accurate code, you should continue the conversation to fine-tune the outcome.

Summary

In this chapter, we discussed the benefits of AI and its application in IoT projects, explored various effective prompting frameworks, and provided an IoT project example of instructing ChatGPT to generate code on ESP32.

In the next chapter, we will review the top 10 popular IoT project examples currently in the market and provide corresponding coding prompts.

7

Recommendations to Start Your First IoT Project

Hello and welcome to *Chapter 7*!

In the last chapter, we covered the introduction of AI and ChatGPT prompt frameworks, along with pragmatic skills and tips to apply ChatGPT during the IoT innovation journey. With these tools, you have what it takes to bring your innovative ideas to life. Congratulations! You are now ready and eager to embark on your first IoT project. This chapter aims to provide practical recommendations to navigate the initial steps of your project smoothly and to help you sidestep potential frustrations.

In this chapter, we'll cover the following topics:

- **Thinking big and starting small**: Embrace your grand ideas for IoT innovation with excitement, but break down your project into small, manageable, and granular phases. This way, you can keep your passion going and positively learn as you progress, without getting overwhelmed by your big plan right from the beginning.

- **Reaching out for the low-hanging fruits first**: Some projects may have less complexity, while others can be more demanding. As a beginner, it is smart to begin with simpler tasks for your first attempt. This will allow you to gradually build your knowledge and skills before taking on more challenging projects.

Technical requirements

The following are the most well-recognized, user-friendly, and affordable starter kits for beginners entering the world of IoT with minimal investment:

- Microsoft Visual Studio Code with the PlatformIO extension
- ChatGPT
- ESP32 MCU

- Arduino-compatible sensors
- AWS Free Tier cloud services

Thinking big and starting small

Imagine you are very excited about creating a smart irrigation solution with a soil moisture measurement sensor, a wildfire sentry solution with a flame/smoke detection sensor, a river/lake pollution monitoring system with a water quality measurement sensor, a CO_2 emission warning system on top of a chimney with an air quality measurement sensor, a roof safety monitoring system with a vibration and tilt sensor, an alert system with ultrasound and body **PIR (Passive Infrared)** sensors when people are approaching a hazardous zone, a water leakage report solution in a restroom through a water drop sensor, or even a mouse trap connected to the cloud to trigger an alarm on your mobile phone. Those grand ideas are motivating you to roll up your sleeves and develop an IoT innovation from scratch, moving to a fully functional solution as swiftly as possible.

Thinking bigger is always encouraged by any means. By thinking bigger, we unleash our potential to dream beyond what seems possible and envision a future that is bold and inspiring. It pushes us to explore new ideas, embrace innovation, and strive for excellence. However, when it comes to practical aspects, it's ideal to switch to a different approach to ensure the holistic development of your project.

As we discussed in previous chapters, a fully functional IoT solution usually includes end devices for indoor or outdoor use, wireless connectivity within a residential room or across a wide area in a city or rural setting, and data processing on a cloud platform that is customized to meet different customer needs. Building such a solution from day one is undoubtedly a challenging task.

Starting small is an advocated practical strategy in this context. It involves breaking down the final deliverable into smaller and gradual phases. By taking practical steps and gradually building knowledge, we can accumulate valuable experience along the way. This approach allows us to make progress, learn from our mistakes, and adjust our course of action as necessary.

For instance, if you are planning to invent a wildfire sentry solution, the fully functional solution includes an outdoor device equipped with flame/smoke sensors and a long-range wireless backhaul. This solution is targeted for deployment in rural areas such as forests or farm fields. The top requirement for such a device is low power consumption. For example, it should be able to operate for at least one year using four AA batteries. Additionally, you must consider the availability of wireless coverage in those areas. Usually, NB-IOT, LTE-M, or LoRaWAN are suitable options.

By adhering to the *start small* philosophy, during the initial concept validation phase, you can begin with an ESP32 with some Arduino-compatible sensors. You can use its built-in Wi-Fi function to access your home Wi-Fi, allowing you to test your idea through rapid prototyping. If the idea works, you can progress to the next step and switch to an ESP32 module with an NB-IOT/LTE-M radio or LoRaWAN radio. This will allow you to evaluate their power consumption and set the stage for further development based on real-world performance.

Applying a *think big and start small* philosophy means setting grand visions while also focusing on achievable beginnings. This sets a solid foundation for larger goals, leading to a *reach out for the low-hanging fruits first* strategy. By capitalizing on easy opportunities, we gain quick wins and momentum, enabling us to tackle more complex challenges and turn our big thinking into actionable paths.

Reaching out for the low-hanging fruits first

The complexity of IoT projects varies widely, from those that demand minimal effort to those demanding domain expertise. For beginners, it is advisable to initially focus on simpler tasks. This strategy enables a gradual enhancement of your knowledge and skills, laying a solid foundation that prepares you to tackle more complex projects in the future.

The following table presents the focus of projects at different complexity levels.

Project complexity level	Entry (beginner-friendly)	Medium (experienced player)	Advanced (domain expertise)
Service availibility	Tolerate occasional errors and temporary service interruptions without significant consequences	Utilize re-transmission mechanisms at the application layer if there are errors, and ensure data integrity with checksums or hashes	Employ advanced error correction technologies, ensure high availability and fault tolerance, and establish stringent SLAs to handle critical data with minimal downtime
Deployment environment	Residential (indoor)	Campus or commercial buildings (indoor and outdoor) and controlled environments	Varied environments, including cities, rural areas, farms, forests, and frontiers with exposure to harsh weather and conditions, requiring ruggedized equipment
End-device physics	Consumer grade	Commercial grade, designed for sustained use and moderate environmental variation	Industrial grade, built to withstand extreme temperatures, moisture, UV rays from the sun, and strong gusts
Wireless connectivity options	Home Wi-Fi or BLE hub	Enterprise Wi-Fi and LoRaWAN, offering greater coverage and penetration through obstacles	NB-IoT, LTE-M, or LoRaWAN, optimized for long-range communication, penetration, and power efficiency

Project complexity level	Entry (beginner-friendly)	Medium (experienced player)	Advanced (domain expertise)
End-device roaming	Fixed on a wall or ceiling	Fixed or local area roaming (allows a service continuity at local area)	Fixed or wide area roaming (demands seamless connectivity)
End-device power supply	External source	Batteries or an external source, with a focus on energy efficiency	Batteries or renewable energy sources such as solar panels, prioritizing self-sufficiency and sustainability, potentially coupled with energy-harvesting technologies
Data communication with the cloud	Primarily upstream, rarely receiving data from the cloud	Bidirectional communication, downstream for unicast, and supporting command and control	Robust bidirectional communication, downstream for broadcast or multicast, supporting complex interactions such as **OTA (Over The Air)** updates, remote diagnostics, and control
Mobile device involvement	SMS or email notification	SMS or email notification	Comprehensive mobile app integration for real-time monitoring and control, with support for notifications, automation rules, and user customization
Application examples	Indoor temperature and humidity monitor, air quality monitor, connected mouse traps, smart lighting, smart doorbells, smart cameras, smoke detectors, and so on	Smart parking, smart lighting, smart cameras, asset tracking, facility security monitoring, HVAC automation, water leakage, smart irrigation, smoke detectors, and so on	Water and gas metering, street lighting, noise detection, waste collection, fleet, agriculture, natural disaster forecast, wildfire alerting, water and air quality monitoring, soil moisture measurement, asset tracking, wild animal tracking, and so on

In this section, we provided a structured overview of the different complexities of IoT projects. This will help you assess and choose projects that match your current expertise level. Starting with beginner-friendly projects can help solidify your foundational knowledge and build confidence. It also prepares you for gradual advancement into more sophisticated applications. Use this guide to strategically plan your IoT learning journey, making sure that each step is meaningful and contributes to your overall mastery of the field.

Summary

In this chapter, we covered the *think big and start small* philosophy, highlighting the importance of making progress step by step in the large and complex world of IoT. You've learned the value of being patient and the strategic method of starting with easy tasks – beginning with projects that align with your growing skills to establish a strong base of knowledge and experience.

In the next chapter, you will find a collection of project samples that are designed for beginners and carefully selected to enhance your learning and inspire your creativity. These projects serve as stepping stones, putting into practice the principles we have discussed and giving you the opportunity to bring your ideas to life. With a solid foundation, it's time to get hands-on and dive into the practical experience that will transform your understanding of IoT.

8

10 Beginner-Friendly IoT Projects with ChatGPT Prompts

Welcome to the innovation workshop, where your ideas will come to life as you build, test, and refine real-world IoT solutions.

In the preceding chapter, we explored foundational advice for beginners embarking on their first IoT projects, emphasizing the philosophy of *Thinking big and starting small*, and *Reaching out for the low-hanging fruits first*.

This chapter is designed to help you transition from theory to practice. We will explore 10 beginner-friendly IoT project examples that showcase a range of applications. Through these examples, you will learn about the integration and interaction of diverse sensors with **ESP32**, and how these components work together to collect data. Each project is carefully selected to comply with the philosophy we've talked about in the previous chapter, *allowing* you to build a solid understanding and gain confidence as you progress.

The 10 examples cover the following projects:

- Project 1 – temperature and humidity measurement
- Project 2 – flame detection
- Project 3 – PIR motion detection
- Project 4 – gas detection
- Project 5 – distance measurement
- Project 6 – tilt detection
- Project 7 – vibration detection
- Project 8 – collision detection
- Project 9 – soil moisture detection
- Project 10 – magnetic change detection

Technical requirements

In the following 10 project cases, we are going to select ESP32-C3 as an example of interaction with the Arduino sensors. The hardware specification and pin definition of ESP32-C3 are illustrated in *Chapter 11*.

Here is the list of sensors covered in this chapter:

Projects	Sensor Module
Temperature and humidity measurement	DHT11
Flame detection	KY-026
PIR motion detection	HC-SR501
Gas detection	MQ-2
Distance measurement	HC-SR04
Tilt detection	SW-520D
Vibration detection	SW-420
Collision detection	Switch on/off sensor
Soil moisture detection	HS-S09
Magnetic change detection	KY-003

Project 1 – temperature and humidity measurement

This project makes use of a DHT11 sensor, a digital sensor that is both compact and affordable. The DHT11 sensor is commonly utilized in a diverse range of applications. These applications range from hobbyist projects created by individuals in their personal time, to larger-scale consumer electronics that are widely available on the market. The DHT11 sensor is a versatile device that combines multiple functions into one compact unit: it contains a thermistor for the purpose of measuring temperature and a capacitive humidity sensor to gauge humidity levels. These two components work in conjunction to provide comprehensive environmental data. Moreover, the DHT11 sensor is also equipped with a high-performance 8-bit microcontroller. This microcontroller enables the DHT11 sensor to offer reliable and accurate readings. Furthermore, the DHT11 sensor presents these readings via a simple digital interface, making it an accessible option for many users, regardless of their level of technical proficiency.

Specifications

Specification	Description
Interface	3 pins, VCC, GND, and signal output
Operating voltage	3.3 V to 5 V
Operating current	2.5 mA on measuring, 60 uA on standby

Specification	Description
Operating temperature	0 °C to 50 °C
Temperature measuring range	0 °C to 50 °C (32 °F to 122 °F) with a ±2 °C accuracy
Humidity measuring range	20 % to 90 % **relative humidity** (**RH**) with ±5 % RH accuracy
Resolution	Temperature: 1 °C; humidity: 1 % RH
Sampling rate	Once every second
Arduino library	DHT.h

Applications

- **Home environment monitoring**: For DIY weather stations or indoor climate monitoring systems

- **Educational projects**: Ideal for teaching the basics of sensor integration and environmental monitoring

- **Agriculture and gardening**: To monitor greenhouse or soil conditions

- **Home automation**: Can be integrated into systems for controlling **heating, ventilation, and air conditioning** (**HVAC**) based on temperature and humidity levels

Prompt to ChatGPT

Role:

Act as a senior software developer specializing in embedded C++ development. Your expertise includes IoT projects using the ESP32 chip, Arduino-compatible sensors, and the PlatformIO IDE.

Task:

As a mentor who educates a high school student and has a basic understanding of Python but is new to C++, your C++ code snippet should be instructional and meet the specified objectives and requirements, catering to a beginner's level.

Objective:

Create a clear and instructive code example for ESP32 C3 that accomplishes the following:

- Reads the temperature and humidity values from the DHT11 sensor every 3 seconds, and compares the values to a predefined range

- Beeps the piezo buzzer and changes the LED color to indicate an abnormal situation

Requirements:

- Use PlatformIO as your IDE

- Connect ESP32 GPIO 0 to the signal output pin of DHT11

- Connect ESP32 GPIO 2, 3, 10 to the R/G/B pins of a 3-color LED

- Connect ESP32 GPIO 11 to the signal pin of the piezo buzzer

- Use Pulse-Width Modulation (PWM) to control the piezo buzzer and the three-color LED

- The LED turns to a steady green color when the temperature falls between 15°C to 30°C and humidity between 10 % to 80 %

- The LED turns to a steady red color when the temperature exceeds 30 °C or the humidity is over 80 %, and beeps fast-paced, high-pitch buzzer alerts

- The LED turns to a steady blue color when the temperature is below 15 °C or humidity is less than 10 %, and beeps fast-paced, high-pitch buzzer alerts

- Print the current value temperature, humidity value, LED color, and beep on/off status

This is what we expect from the output:

- Structure your output code snippets in the following order: import libraries, constants and variable declaration, hardware initial settings, functions initial declaration, setup function, main loop function, and functions definition.

- Provide detailed comments to each code line for clarity.

- Apply C++ programming best practices to this code snippet complying with the following rules:

  - Code simplicity: Apply readability and simplicity over-complicated solutions. Refactor large functions into smaller, more manageable pieces.

  - Consistency: Adhere to a consistent code style in your project, such as indentation, bracket placement, and whitespace usage.

  - Naming conventions: Use camelCase for variables and functions definition, and PascalCase for class names.

- Modularization: Split the code into logical modules or components, such as function or class, each handling a specific part of the program's functionality.

- Error handling: Ensure exception safety and check the return value.

Code example

The `main.cpp` example code is at the following link: `https://github.com/PacktPublishing/ Accelerating-IoT-Development-with-ChatGPT/blob/main/Chapter_8/ Project_1/main.cpp`.

In this example code, as we instructed, ChatGPT created the following four functions:

- `void readDHTSensor()`: This reads the sensor's output data, such as temperature in Celsius and Fahrenheit and humidity

- `void updateIndicatorStatus()`: This changes the color of the LED and calls the `Buzzer` beep function according to the comparison of actual data to the predefined data range

- `void beepBuzzerAlert()`: Beeps the on/off buzzer

- `void printSystemStatus()`: This prints out the current sensor data, the LED's color, and the buzzer on/off status in each loop

In *Chapter 11*, there is an instruction to use PlatformIO to compile and upload code in ESP32; you can follow that instruction to validate this example code to observe if these four functions fully meet our requirements.

In this project, we periodically read data from the DHT11 sensor to monitor temperature and humidity data. By comparing these readings to predefined ranges, we utilize an RGB LED and a piezo buzzer to visually and audibly indicate the status. This process helps in understanding how to handle sensor data and respond to different conditions using basic components and C++ programming.

Building on this foundation, our next project will focus on flame detection. We will design a similar code structure to read digital values from a flame sensor pin and apply the same approach to indicate the presence of flame using LEDs and a buzzer. This will further enhance your skills in integrating sensors and actuators in embedded systems while reinforcing the principles of condition-based responses in IoT applications.

Project 2 – flame detection

This project utilizes the KY-026 module, a sophisticated flame detection module specifically designed for use with Arduino and a variety of other microcontrollers. The primary function of this module is to detect the infrared light that is typically emitted by fire.

At the heart of the KY-026 module is the YG1006 sensor. This is a **negative-positive-negative (NPN)** silicon phototransistor that is known for its high-speed performance and remarkable sensitivity. These characteristics allow it to rapidly and accurately detect the presence of fire or excessive heat.

One of the key features of this module is its dual output capability, offering both digital and analog outputs. This dual function allows users to select the most appropriate output for their specific application, providing a high level of flexibility in usage.

The KY-026 module finds applications in an impressively wide range of scenarios that require fire or heat detection. For instance, it can be integrated into safety systems in residential environments, providing an additional level of security against fire hazards. In industrial settings, it can be used to monitor and control equipment that operates at high temperatures, thus preventing potential accidents.

Specifications

Specification	Description
Interface	4 pins, VCC, GND, analog output (A0) and digital output (DO)
Output type	Analog output provides a continuous voltage that varies with the flame's intensity, and digital output provides LOW for no flame detected, and HIGH for a flame detected
Operating voltage	3.3 V to 5 V
Operating current	Very low
Operating temperature	-25 °C to +85 °C
Detection wavelength	760 nm to 1100 nm
Detection angle	60 degrees

Analog output values

No flame detected	The analog value might be close to 1023
Weak flame or far distance	Values might range in the upper mid-range, such as 600-800
Moderate flame or medium distance	Mid-range values, such as 300-600
Strong flame or close proximity	Lower values, potentially below 300, indicate a strong flame or close proximity to the sensor

Applications

- **Fire alarm systems**: For DIY fire alarm systems that trigger an alert or take action when a flame is detected

- **Safety devices**: Integrated into safety devices to provide warnings or shut down systems in the presence of fire

- **Environmental monitoring**: Used in environmental monitoring setups to detect fire outbreaks in monitored areas
- **Robotics**: In robotics projects for navigating or avoiding flames and high-temperature areas

Prompt to ChatGPT

The role, task, and output expectation will be the same as the first project here:

Role:

Act as a senior software developer specializing in embedded C++ development. Your expertise includes IoT projects using the ESP32 chip, Arduino-compatible sensors, and the PlatformIO IDE.

Task:

As a mentor who educates a high school student and has a basic understanding of Python but is new to C++, your C++ code snippet should be instructional and meet the specified objectives and requirements, catering to a beginner's level.

Objective:

Create a clear and instructive code example for ESP32-C3 that accomplishes the following:

- Reads the digital value (HIGH or LOW) from KY-026 every 3 seconds
- Beeps the piezo buzzer and changes the LED color to indicate flame is detected

Requirements:

- Use PlatformIO as your IDE
- Connect ESP32 GPIO 0 to the KY-026 D0 pin
- Connect ESP32 GPIO 2, 3, 10 to the R/G/B pins of a 3-color LED
- Connect ESP32 GPIO 11 to the signal pin of the piezo buzzer
- Use PWM to control the piezo buzzer and the three-color LED
- If the value is HIGH, which indicates the flame is detected, beep the piezo buzzer and change the LED color to red
- If the value is LOW, which indicates no flame is detected, mute the piezo buzzer and maintain the LED color to green
- Print the current value, beep on/off status, and LED color

This is what we expect from the output:

- Structure your output code snippets in the following order: import libraries, constants and variable declaration, hardware initial settings, functions initial declaration, setup function, main loop function, and functions definition.

- Provide detailed comments to each code line for clarity.

- Apply C++ programming best practices to this code snippet complying with the following rules:

  - Code simplicity: Apply readability and simplicity over-complicated solutions. Refactor large functions into smaller, more manageable pieces.

  - Consistency: Adhere to a consistent code style in your project, such as indentation, bracket placement, and whitespace usage.

  - Naming conventions: Use camelCase for variables and functions definition, and PascalCase for class names.

  - Modularization: Split the code into logical modules or components, such as function or class, each handling a specific part of the program's functionality.

  - Error handling: Ensure exception safety and check the return value.

Code example

An example of the main.cpp code can be found at the following link: https://github.com/PacktPublishing/Accelerating-IoT-Development-with-ChatGPT/blob/main/Chapter_8/Project_2/main.cpp.

In this example code, the following three functions were reused from the previous project:

- void updateIndicatorStatus()

- void beepBuzzerAlert()

- void printSystemStatus()

Instead of reading the analog value of the flame sensor, we use a new Boolean function, isFlameOn(), to read the digital value (LOW for no flame detected or HIGH for flame detected), and return the comparison result as true or false, simplifying the logic for determining the system's response to the flame sensor's output.

In this project, we effectively determined the presence of a flame by examining the digital value of the flame sensor pin in each loop period. This approach allowed us to use an LED and a buzzer to indicate whether a flame was detected, reinforcing our understanding of condition-based responses using digital sensor readings.

Building on this knowledge, our next project will involve using a similar approach to read the digital value from a PIR motion sensor. This project aims to detect the presence of people or animals in front of the sensor, further enhancing your skills in integrating sensors for real-time monitoring and alert systems in IoT applications.

Project 3 – PIR motion detection

The project employs the HC-SR501, a PIR sensor that is engineered specifically for the purpose of detecting the presence of humans or animals. It achieves this through the utilization of infrared signals. The sensor offers two distinct operating modes, which are *Repeatable (H)* and *Non-Repeatable (L)*. These modes offer flexibility in terms of application and use. The sensor finds itself particularly well suited for a wide variety of applications, including but not limited to automatic lighting, security alarms, and industrial automation control.

One of the standout features of this sensor is its wide temperature range capabilities. This means it can function effectively in a range of different environmental conditions. In addition to this, it also boasts low power consumption. This makes it an economical choice, particularly for long-term deployment. As such, it proves to be a versatile option for any motion detection requirements, delivering both efficiency and adaptability.

Specifications

Specification	Description
Interface	3 pins, VCC, GND, and signal output (HIGH/LOW)
Output type	HIGH for motion detected and LOW for no motion detected
Operating voltage	4.5 V to 20 V
Operating current	50 µA
Operating temperature	-20 °C to +80 °C
Detection range	Up to 7 meters (about 23 feet)
Viewing angle	Up to 100 degrees
Output delay time	Adjustable from 5 seconds to 200 seconds

Applications

- **Security systems**: For motion-based security devices, triggering alarms or lights upon detection of an intruder

- **Automatic lighting**: To turn on lights in a room or area when someone enters and turn off when left unoccupied

- **Home automation**: Integrated into smart home systems for energy-saving and convenience features

- **Wildlife monitoring**: Used in wildlife cameras to capture images or videos when motion is detected

Prompt to ChatGPT

The role, task, and output expectation will be the same as the first project here. The other aspects are as follows:

```
Objective:

Create a clear and instructive code example for ESP32-C3 that
accomplishes the following:
```

- Reads the digital value (HIGH or LOW) from HC-SR501 every 3 seconds
- Beeps the piezo buzzer and changes the LED color to indicate motion is detected

```
Requirements:
```

- Use PlatformIO as your IDE
- Connect ESP32 GPIO 0 to the signal out output pin of HC-SR501
- Connect ESP32 GPIO 2, 3, 10 to the R/G/B pins of a 3-color LED
- Connect ESP32 GPIO 11 to the signal pin of the piezo buzzer
- Use PWM to control the piezo buzzer and the three-color LED
- If the value is HIGH, which indicates a motion is detected, beep the piezo buzzer and change the LED color to red
- If the value is LOW, which indicates no motion is detected, mute the piezo buzzer and maintain the LED color to green
- Print the current value, beep on/off status, and LED color

Code example

An example of the `main.cpp` code can be found at the following link: `https://github.com/PacktPublishing/Accelerating-IoT-Development-with-ChatGPT/blob/main/Chapter_8/Project_3/main.cpp`.

In this example code, we continue to use the same three functions from *Projects 1* and *2*:

- `void updateIndicatorStatus()`
- `void beepBuzzerAlert()`
- `void printSystemStatus()`

Additionally, we introduce a Boolean function, `isPIROn()`, to read the digital value (`HIGH` for motion detected or `LOW` for no motion detected) from the PIR sensor. This function returns the comparison result as `true` or `false`, simplifying the logic for determining the system's response to motion detection.

In this project, we successfully used the PIR sensor to detect motion by examining its digital output. This allowed us to indicate motion detection through an LED and a buzzer, reinforcing our ability to handle digital sensor readings and implement responsive alerts.

In the next project, we will extend this approach to a gas sensor. By reading the digital value from the gas sensor, we will design a system to alert us if any gas is detected, continuing to build your expertise in real-time monitoring and alert systems using various sensors and ESP32.

Project 4 – gas detection

This project adopts an MQ-2 sensor, a sensor that has garnered widespread acceptance and use in the field of gas detection equipment due to its robust capabilities. The MQ-2 sensor is not only capable but also highly proficient in detecting an extensive variety of gases. These include **liquefied petroleum gas (LPG)**, i-butane, propane, methane, alcohol, hydrogen, and even smoke.

This wide range of detection capabilities makes the MQ-2 sensor an incredibly versatile tool. It's not just the variety of gases the sensor can detect that separates it from the rest, but also its speed and sensitivity. The sensor is highly regarded for its rapid response time, a feature that is particularly valuable in critical situations where time is of the essence. Furthermore, its high sensitivity allows for the detection of even minute amounts of gas, which can be crucial in preventing potential hazards.

Given these unique attributes, the MQ-2 sensor is an ideal choice for myriad safety applications, especially those where the detection of gas leakage is of paramount importance.

Specifications

Specification	Description
Interface	4 pins, VCC, GND, analog output (A0) and digital output (D0)
Output type	Analog voltage output proportional to gas concentration and digital output by HIGH for gas detection and LOW for no gas detection
Operating voltage	5 V
Operating temperature	-20 °C to 50 °C
Target gases	LPG, i-butane, propane, methane, alcohol, hydrogen, and smoke
Detection range	Approximately 300 to 10,000 **parts per million (ppm)** for various gases
Preheat duration	Around 20 seconds to 1 minute for initial heating
Load resistance	Adjustable, typically around 5 KΩ
Heater consumption	Approximately 800 mW

Analog output values

Clean air	The output voltage in clean air (baseline) might be around 1 V to 1.5 V.
Low gas concentration	A slight increase in gas concentration might increase the voltage slightly above the baseline level. For instance, detecting low levels of methane might result in output voltages in the range of 1.5 V to 2.5 V.
High gas concentration	In the presence of high concentrations of combustible gas, the output voltage could approach the maximum operating voltage of the sensor (close to 5 V for a 5V supply).

Applications

- **Gas leak detectors**: For residential or commercial environments to detect dangerous gas leaks
- **Air quality monitoring**: To monitor the presence and concentration of various gases in the air
- **DIY projects**: Popular in DIY electronics projects involving gas detection and environmental monitoring
- **Safety alarms**: Integrated into safety systems to trigger alarms in the presence of combustible gases

Prompt to ChatGPT

The role, task, and output expectation will be the same as the first project here. The other aspects are as follows:

Objective:

Create a clear and instructive code example for ESP32-C3 that accomplishes the following:

- Reads the digital value (HIGH or LOW) from MQ-2 every 3 seconds
- Beeps the piezo buzzer and changes the LED color to indicate gas is detected

Requirements:

- Use PlatformIO as your IDE
- Connect ESP32 GPIO 0 to the D0 pin of MQ-2
- Connect ESP32 GPIO 2, 3, 10 to the R/G/B pins of a 3-color LED
- Connect ESP32 GPIO 11 to the input pin of the piezo buzzer
- Use PWM to control the piezo buzzer and the three-color LED
- If the value is HIGH, which indicates gas is detected, beep the piezo buzzer and change the LED color to red
- If the value is LOW, which indicates no gas is detected, mute the piezo buzzer and maintain the LED color to green
- Print the current value, beep on/off status, and LED color

Code example

An example of the main.cpp code can be found at the following link: https://github.com/PacktPublishing/Accelerating-IoT-Development-with-ChatGPT/blob/main/Chapter_8/Project_4/main.cpp.

In this example code, we follow the same function structure as in *Project 3*. The isGasOn() Boolean function is used to check for the presence of gas by reading the digital value (HIGH for gas detected or LOW for no gas detected) from the MQ-2 sensor. This function returns a Boolean value to simplify the logic for determining the system's response to gas detection.

In this project, we used the digital output of the MQ-2 sensor to detect various gases. This enabled us to signal gas detection with an LED and a buzzer, honing our skills in managing digital sensor readings and setting up responsive alerts.

For our next project, we plan to use the HC-SR04 ultrasonic sensor to gauge the distance to objects by leveraging its transmission and receiving antennas. This project will deepen our understanding of sensor integration and real-time data processing in embedded systems.

Project 5 – distance measurement

This project prominently features the HC-SR04 sensor, a commonly used ultrasonic sensor in a multitude of Arduino projects, especially those involving the measurement of distance. The HC-SR04 ultrasonic sensor operates on a basic yet effective principle. It emits an ultrasonic wave, a type of sound wave, at a frequency of 40 kHz. This wave travels seamlessly through the air, functioning as the sensor's probing mechanism. If there happens to be an object or an obstacle in the path of this emitted wave, the sound wave is reflected back, effectively bouncing back to the sensor. This phenomenon is akin to the echo one might hear in a large, empty room or a mountain valley. By measuring the time interval between the emission of the sound wave and the reception of the echo, the sensor can accurately calculate the distance to the object or obstacle. This precise and reliable functionality has made the HC-SR04 ultrasonic sensor a popular choice for a wide variety of applications. These include but are not limited to robotics, obstacle avoidance systems, and range detection projects.

Specifications

Specification	Description
Interface	4 pins, VCC, GND, Trig (trigger), and Echo (receive)
Trigger input signal	10 μS **TTL (Transistor-Transistor Logic)** pulse
Echo output signal	Input TTL level signal and the duration is proportional to the range
Operating voltage	5 V DC
Operating temperature	-20 °C to +70 °C
Operating current	15 mA
Ultrasonic frequency	40 kHz
Max range	4 m (about 13 feet)
Min range	2 cm (about 0.8 inches)
Measuring angle	15 degrees
Resolution	0.3 cm

Applications

- **Obstacle avoidance systems**: In robotics, the HC-SR04 sensor is used to detect obstacles in the path of a robot, allowing it to navigate around them
- **Distance measurement**: The sensor is used in applications requiring precise distance measurements, such as in manufacturing and quality control processes

- **Parking sensors**: In automotive applications, the HC-SR04 can be used to develop parking assistance systems to alert drivers about the distance to objects behind the vehicle

- **Security and surveillance**: The sensor can be integrated into security systems to detect the presence of persons or objects in restricted areas

Prompt to ChatGPT

The role and task will be the same as the first project here. The other aspects are as follows:

```
Objective:
```

```
Create a clear and instructive code example for ESP32-C3 that
accomplishes the following:
```

- Reads the output values from HC-SR04 every 100 milliseconds
- Beeps the piezo buzzer and changes the LED color to indicate an abnormal distance range

```
Requirements:
```

- Use PlatformIO as your IDE
- Connect ESP32 GPIO 0 to the Trig pin and pin 1 to the Echo pin of HC-SR04
- Connect ESP32 GPIO 2, 3, 10 to the R/G/B pins of a 3-color LED
- Connect ESP32 GPIO 11 to the signal pin of the piezo buzzer
- Use PWM to control the piezo buzzer and the three-color LED
- If the distance value is within 10 cm, beep the piezo buzzer and change the LED color to red
- If the distance value is between 10 cm to 30 cm, beep the piezo buzzer and change the LED color to blue
- If the distance value is beyond 30 cm, mute the piezo buzzer and maintain the LED color to green
- Print the current value, beep on/off status, and LED color

Code example

An example of the main.cpp code can be found at the following link: https://github.com/PacktPublishing/Accelerating-IoT-Development-with-ChatGPT/blob/main/Chapter_8/Project_5/main.cpp.

In this project, we utilized the HC-SR04 ultrasonic sensor to report real-time distance measurements from the sensor to objects in its path. Unlike our previous projects, which returned Boolean results, indicating the presence or absence of a condition, this project uses the `readUltrasoundSensor()` function to calculate and return the measured distance. This distance is calculated by measuring the time lapse between sending a pulse from the Trig pin and receiving the echo at the Echo pin, using the `SOUND_SPEED_CM_PER_US = 0.017` constant, which represents the speed of sound in cm/µs, divided by 2. The distance thus measured is indicated using an LED and a buzzer, enhancing our proficiency in processing real-time sensor data and implementing responsive alert systems.

For our next project, we plan to incorporate the SW-520D sensor for tilt detection. This project will focus on enhancing our understanding of monitoring tilt status changes and responding promptly. This will further cultivate our expertise in integrating diverse sensor types for real-time monitoring and alert systems.

Project 6 – tilt detection

This project employs the use of the SW-520D sensor, a simplistic yet highly effective tilt sensor that boasts a wide range of applications. This versatility makes it suitable for anything from small-scale hobbyist projects to more complex and advanced electronic systems. The core of the sensor is comprised of two metal balls housed within a cylindrical case.

The functionality of this sensor is based on its position relative to the ground. When the sensor is tilted beyond a 15-degree angle from a horizontal position, the two metal balls make contact with the terminals, subsequently closing the circuit. This mechanism allows the sensor to detect significant shifts in orientation.

However, when the sensor returns to a horizontal state or is tilted less than 15 degrees, the metal balls will not touch the terminals. This results in the circuit remaining open.

The practical applications of the SW-520D sensor are numerous. By detecting the orientation or movement of an object, it can be used in various contexts such as monitoring the opening and closing of a door, tracking the movement of a robot, or even as a component in a toy.

Specifications

Specification	Description
Interface	3 pins, VCC, GND, digital output (HIGH/LOW)
Output type	HIGH for tilt detected, and LOW upon no tilt detected
Operating voltage	3.3 V to 5 V
Operating current	15 mA
Operating temperature	-20 °C to +80 °C
Sensing angle	15 to 80 degrees from the horizontal typically triggers change

Applications

- **Alarm systems**: Used in security devices to detect tampering or unauthorized movement of objects, such as safes, doors, or windows

- **Smart home devices**: Used in smart lighting systems to turn lights on/off or adjust brightness based on the tilt angle, or in smart window treatments to adjust blinds or curtains automatically

- **Leveling instruments**: Used in construction and surveying equipment to ensure level measurements and alignments are accurate

- **Personal safety devices**: Incorporated into elderly care products, such as fall detectors that alert caregivers when a fall is detected

Prompt to ChatGPT

The role, task, and output expectation will be the same as the first project here. The other aspects are as follows:

```
Objective:

Create a clear and instructive code example for ESP32-C3 that
accomplishes the following:
```

- Reads the digital value (HIGH or LOW) from SW-520D every 3 seconds
- Beeps the piezo buzzer and changes the LED color to indicate whether tilt is detected

```
Requirements:
```

- Use PlatformIO as your IDE
- Connect ESP32 GPIO 0 to the D0 pin of SW-520D
- Connect ESP32 GPIO 2, 3, 10 to the R/G/B pins of a 3-color LED
- Connect ESP32 GPIO 11 to the signal pin of the piezo buzzer
- Use PWM to control the piezo buzzer and the three-color LED
- If the value is HIGH, which indicates tilt is detected, beep the piezo buzzer and change the LED color to red
- If the value is LOW, which indicates no tilt is detected, mute the piezo buzzer and maintain the LED color to green
- Print the current value, beep on/off status, and LED color

Code example

At `https://github.com/PacktPublishing/Accelerating-IoT-Development-with-ChatGPT/blob/main/Chapter_8/Project_6/main.cpp`, we provide a piece of example code. The code structure is exactly the same as in *Projects 2, 3,* and *4.* It uses an `isTiltOn()` Boolean function to read the digital value from the tilt sensor pin and then returns the result as true or false.

In this project, we utilized the SW-520D tilt sensor to monitor tilt changes. We accomplished this by reading its digital pin and interpreting the data. This approach enabled us to detect tilt events and respond appropriately, thereby improving our knowledge of condition-based responses with digital sensors.

For the next project, we aim to use the SW-420D sensor for vibration detection. We plan to use a similar approach by sensing the vibration status via digital pin readings. This will further enhance our capabilities in integrating various sensor types for real-time monitoring and alert systems.

Project 7 – vibration detection

This project uses SW-420, a highly sensitive vibration sensor module that has been meticulously designed to detect even the slightest movements.

From being an integral part of high-tech security systems that safeguard our homes and businesses to playing a crucial role in advanced alarm systems that alert us of potential dangers, the SW-420 sensor has proven its versatility. It's not just limited to these areas; the sensor also finds its place in smart vehicles, contributing to safer and smarter transportation.

Furthermore, in the realm of natural disaster management, it is instrumental in earthquake detection systems, providing early warnings to prevent loss of life and property. The SW-420 sensor is more than a sensor; it's a life-saving technology.

One of the crowning features of this module is its vibration sensor component. This component is designed to generate a signal as soon as the detected vibrations cross a predetermined threshold. This unique feature makes the SW-420 sensor perfect for detecting unauthorized or unexpected movements, adding a layer of security and control to any system it's integrated into.

Specifications

Specification	Description
Interface	3 pins, VCC, GND, digital output (`HIGH`/`LOW`)
Output type	`HIGH` for vibration detected, and `LOW` for no vibration detected
Operating voltage	3.3 V to 5 V
Operating current	<15 mA
Operating temperature	-20 °C to +70 °C
Sensing sensitivity	Adjustable via onboard potentiometer

Applications

- **Security systems**: For detecting unauthorized access through doors or windows

- **Vehicle detection**: This can be used to detect vibrations caused by vehicles moving nearby

- **Earthquake monitoring**: In DIY projects for detecting tremors and vibrations

- **Industrial monitoring**: To monitor machinery for unexpected vibrations that could indicate malfunctions or maintenance needs

Prompt to ChatGPT

The role, task, and output expectation will be the same as the first project of this chapter. The other aspects are as follows:

Objective:

Create a clear and instructive code example for ESP32-C3 that accomplishes the following:

- Reads the digital value (HIGH or LOW) from SW-420 every 1 second

- Beeps the piezo buzzer and changes the LED color to indicate whether vibration is detected

Requirements:

- Use PlatformIO as your IDE

- Connect ESP32 GPIO 0 to the D0 pin of SW-420

- Connect ESP32 GPIO 2, 3, 10 to the R/G/B pins of a 3-color LED

- Connect ESP32 GPIO 11 to the signal pin of the piezo buzzer

- Use PWM to control the piezo buzzer and the three-color LED

- If the value is HIGH, which indicates vibration is detected, beep the piezo buzzer and change the LED color to red

- If the value is LOW, which indicates no vibration is detected, mute the piezo buzzer and maintain the LED color to green

- Print the current value, beep on/off status, and LED color

Code example

At `https://github.com/PacktPublishing/Accelerating-IoT-Development-with-ChatGPT/blob/main/Chapter_8/Project_7/main.cpp`, you can find example code. The structure is exactly the same as in *Project 5*. It uses an `isVibrationOn()` Boolean function to read the digital value from the vibration sensor pin and then returns the result as `true` or `false`.

In this project, we used the SW-420 vibration sensor to detect vibration status changes by reading its digital pin and interpreting the result. This method enabled us to detect vibrations and respond accordingly, further enhancing our understanding of condition-based responses using digital sensors.

In the next project, we will use a switch sensor to detect collisions. This project will continue to build your skills in integrating various sensors for real-time monitoring and alert systems, using similar methods to sense and respond to collision events.

Project 8 – collision detection

This project focuses on the concept of collision detection, an essential component in many fields and applications. The main feature of this project is the use of a switch sensor, which is also commonly known as a **bump switch**, a **mechanical switch**, or a **tactile switch**.

These types of switches are simple, yet highly effective, making them a popular choice for detecting physical contact or impact. They are widely used in various fields and applications, ranging from complex robotic systems to critical safety devices. The primary function of these switches is their ability to change states – from `ON` to `OFF` or vice versa – upon encountering any sort of physical force or obstruction.

This unique characteristic makes them exceptionally suitable for tasks requiring basic collision or proximity detection. Whether it is to prevent a robot from bumping into an obstacle or to trigger an alarm when a certain boundary is breached, these switches prove to be highly useful and reliable. The project aims to provide a comprehensive understanding of how to best use these switches in various practical scenarios.

Specifications

Specification	Description
Interface	3 pins, VCC, GND, digital output (`HIGH`/`LOW`)
Output type	`HIGH` for no collision detected, `LOW` for physical contact or force.
Operating voltage	3.3 V to 5 V
Operating temperature	-10 °C to +70 °C

Applications

- **Robotics**: Used in robots as bump sensors to detect obstacles and prevent collisions by altering the robot's path

- **Security systems**: Integrated into security devices as tamper switches to detect unauthorized access or breaches

- **Consumer products**: Employed in toys and consumer electronics as basic input devices or to detect when parts are connected or assembled

- **Industrial machinery**: Used as limit switches to detect the presence or position of machine components for safety and control

Prompt to ChatGPT

The role, task, and output expectation will be the same as the first project here. The other aspects are as follows:

```
Objective:

Create a clear and instructive code example for ESP32-C3 that
accomplishes the following:
```

- Reads the digital value (HIGH or LOW) from the switch sensor every 1 second

- Beeps the piezo buzzer and changes the LED color to indicate whether a collision is detected

```
Requirements:
```

- Use PlatformIO as your IDE

- Connect ESP32 GPIO 0 to the D0 pin of the switch sensor

- Connect ESP32 GPIO 2, 3, 10 to the R/G/B pins of a 3-color LED

- Connect ESP32 GPIO 11 to the signal pin of the piezo buzzer

- Use PWM to control the piezo buzzer and the three-color LED

- If the value is LOW, which indicates collision is detected, beep the piezo buzzer and change the LED color to red

- If the value is HIGH, which indicates no collision is detected, mute the piezo buzzer and maintain the LED color to green

- Print the current value, beep on/off status, and LED color

Code example

Visit our GitHub repository at `https://github.com/PacktPublishing/Accelerating-IoT-Development-with-ChatGPT/blob/main/Chapter_8/Project_8/main.cpp` for an illustrative code sample. In this example, we use the `isCollisionOn()` Boolean function to read the digital value. This function returns a Boolean: `true` for collision detected (when the digital value is `LOW`) and `false` for no collision (when the digital value is `HIGH`). This is a slight variation from previous projects.

In this project, we adopted a switch sensor to report collision status changes by reading its digital pin and interpreting the result. This method allowed us to detect collisions and respond accordingly, further enhancing our skills in condition-based responses using digital sensors.

In the next project, we will concentrate on monitoring soil moisture levels. We will integrate a soil moisture sensor to measure and report soil moisture, furthering our expertise in real-time environmental monitoring and alert systems using various sensors.

Project 9 – soil moisture detection

In this project, we are utilizing a highly specialized soil moisture sensor to gather crucial data about the soil's moisture content. The sensor's mechanism is rooted in the concept of soil resistivity measurement. This technique is implemented by measuring the electrical resistance between two specifically designed probes that are carefully inserted into the soil.

The relationship between the soil's moisture content and its electrical resistance is quite fascinating. Essentially, the higher the moisture content in the soil, the lower the electrical resistance measured between the two probes. This is because water is a good conductor of electricity, and therefore, as the amount of water in the soil increases, it becomes easier for electricity to flow between the probes, thus reducing the measured resistance.

This inverse relationship forms the core working principle of the soil moisture sensor. By constantly monitoring the resistance levels, the sensor accurately gauges the amount of moisture present in the soil at any given moment. This information is critical for a range of applications, from agricultural practices to environmental monitoring and research. It allows for precise irrigation control, helps predict crop yields, and contributes to our understanding of soil health and its impact on plant growth.

Specifications

Specification	Description
PINs	VCC, GND, analog output (A0) and digital output (D0)
Output type	Analog output provides a continuous voltage signal proportional to the sound intensity. Digital output reports HIGH for no moisture detection and LOW for moisture detection

Specification	Description
Operating Voltage	3.3 V to 5.5 V DC
Operating Current	15 mA

Applications

- **Agriculture**:

    - **Irrigation management**: It helps optimize irrigation by providing real-time soil moisture data, ensuring that plants receive the right amount of water

    - **Precision farming**: It allows farmers to monitor soil moisture levels across different parts of a field, leading to more efficient use of water resources

- **Environmental monitoring**:

    - **Soil health assessment**: It monitors soil moisture to assess the health and quality of the soil, which is crucial for sustainable farming practices

    - **Drought monitoring**: It helps in tracking soil moisture trends to identify drought conditions and inform water management strategies

- **Gardening and landscaping**:

    - **Automated irrigation systems**: It integrates with automated watering systems to ensure that gardens and landscapes are watered based on the actual moisture needs of the soil

Prompt to ChatGPT

The role, task , and output expectation will be the same as the first project here. The other aspects are as follows:

```
Objective:

Create a clear and instructive code example for ESP32-C3 that
accomplishes the following:
```

- Reads the digital value (HIGH or LOW) from the moisture sensor every 3 seconds
- Beeps the piezo buzzer and changes the LED color to indicate whether moisture is detected

Requirements:

- Use PlatformIO as your IDE
- Connect ESP32 GPIO 0 to the D0 pin of the moisture sensor
- Connect ESP32 GPIO 2, 3, 10 to the R/G/B pins of a 3-color LED
- Connect ESP32 GPIO 11 to the signal pin of the piezo buzzer
- Use PWM to control the piezo buzzer and the three-color LED
- If the value is LOW, which indicates moisture is detected, beep the piezo buzzer and change the LED color to red
- If the value is HIGH, which indicates no moisture is detected, mute the piezo buzzer and maintain the LED color to green
- Print the current value, beep on/off status, and LED color

Code example

Visit our GitHub repository at https://github.com/PacktPublishing/Accelerating-IoT-Development-with-ChatGPT/blob/main/Chapter_8/Project_9/main.cpp for an illustrative code sample. In this example, we use an isMoistureOn() Boolean function to read the digital value from the soil moisture sensor. A HIGH value indicates no moisture detected, while a LOW value means moisture is detected.

In this project, we used a soil moisture sensor to measure moisture levels in the soil. We read the sensor's digital pin and interpreted the results, which allowed us to monitor soil moisture and respond as needed. This process improved our skills in environmental monitoring and condition-based responses using digital sensors.

In our next project, we will use a magnetic field sensor to detect and report the presence of a magnetic field. This will continue to enhance our expertise in integrating different types of sensors for real-time monitoring and alert systems.

Project 10 – magnetic change detection

This project leverages the capabilities of the KY-003 sensor, a hall magnetic sensor module. This module is a key component in a variety of electronic projects, particularly those that involve Arduino platforms. The main functionality of the KY-003 sensor is to detect magnetic fields. It has a broad range of applications, from identifying the presence or absence of magnetic objects to measuring the exact position of a magnet. The sensor can also be used to count rotational objects when it is combined with magnets, a feature that is highly beneficial in several industrial and scientific contexts.

The KY-003 module incorporates a Hall effect sensor, a device that operates based on a fascinating principle. It generates a voltage difference across an electrical conductor when a magnetic field is applied perpendicular to the flow of current through the sensor. This voltage difference can be detected by a microcontroller, which is often an integral part of the system in which the sensor is embedded.

The presence of a magnetic field is typically indicated by a change in the state of the sensor's digital output pin. Depending on the design of the sensor and the nature of the magnet used, this pin may go HIGH or LOW. This change in state provides a clear, binary signal that can be used to trigger other components or processes within the system.

Specifications

Specification	Description
PINs	3 pins, VCC, GND, and digital output (HIGH/LOW)
Output type	HIGH when it detects a magnetic field and returns to LOW when the field is removed
Operating voltage	5 V
Operating temperature	40 °C to 85 °C
Operating current	Very low

Applications

- **Magnetic field detection**: To sense the presence or absence of a magnetic field

- **Rotational speed measurement**: To count the number of rotations of a magnetic wheel or gear, useful in speedometers and rotational counters

- **Position sensing**: To determine the position of a magnet attached to a moving object, ideal for precision control in robotics

- **Contactless switches**: To create a switch that activates in the presence of a magnetic field, useful in safety and security applications

Prompt to ChatGPT

The role, task, and output expectation will be the same as the first project here. The other aspects are as follows:

```
Objective:

Create a clear and instructive code example for ESP32-C3 that
accomplishes the following:
```

- Reads the digital value (HIGH or LOW) from the hall magnetic sensor every 1 second

- Beeps the piezo buzzer and changes the LED color to indicate whether a magnetic field is detected

Requirements:

- Use PlatformIO as your IDE
- Connect ESP32 GPIO 0 to the D0 pin of the KY-003 sensor
- Connect ESP32 GPIO 2, 3, 10 to the R/G/B pins of a 3-color LED
- Connect ESP32 GPIO 11 to the signal pin of the piezo buzzer
- Use PWM to control the piezo buzzer and the three-color LED
- If the value is HIGH, which indicates that a magnetic field is detected, beep the piezo buzzer and change the LED color to red
- If the value is LOW, which indicates that no magnetic field is detected, mute the piezo buzzer and maintain the LED color to green
- Print the current value, beep on/off status, and LED color

Code example

Visit our GitHub repository at https://github.com/PacktPublishing/Accelerating-IoT-Development-with-ChatGPT/blob/main/Chapter_8/Project_10/main.cpp for an illustrative code sample. In this example, we use the isMagneticOn() Boolean function to read the digital value from the magnetic sensor. A HIGH value indicates a magnetic field is detected, while a LOW value means no magnetic field is detected.

In this project, we utilized a magnetic sensor to detect the presence of a magnetic field by reading its digital pin and interpreting the result. This approach is similar to previous projects involving PIR motion, tilt, and vibration detection, reinforcing your understanding of digital sensor integration and response mechanisms.

After an enriching journey through various project cases, we have now successfully covered all 10 of them. Each project has been meticulously detailed, providing practical examples of code to ensure a comprehensive understanding. These projects range from simple sensor integration to more complex real-time monitoring systems, each offering unique insights and practical experience.

For the best learning experience, we have made available a wealth of resources. You can access detailed code examples on our GitHub repository. This includes the primary files such as main.cpp and platformio.ini, along with prompts to ChatGPT. This repository is neatly organized and can be found at https://github.com/PacktPublishing/Accelerating-IoT-Development-with-ChatGPT/tree/main/Chapter_8.

Summary

The 10 projects in this chapter have been carefully designed to collectively enhance your skills in various key areas. They provide the opportunity to get hands-on with a variety of sensors, each with its unique characteristics and usage. You'll learn how to implement real-time monitoring systems, a crucial aspect of IoT applications. Moreover, these projects will help you understand how to develop responsive IoT systems using ESP32-C3 and PlatformIO. In *Chapter 11*, there are instructions on how to use PlatformIO to compile and upload code in ESP32, and you can apply that instruction to implement these 10 projects.

By working through these projects, you'll not only gain technical skills but also develop problem-solving and critical thinking abilities that are essential in the field of IoT development. We hope this journey has been as enriching for you as it has been for us in curating it.

Starting from the next chapter, we will embark on a practical journey in selecting the DHT11 temperature and humidity sensor. You also can select another case from these 10 projects. That practical journey will begin with learning how to use ChatGPT to draw an IoT application diagram before wiring the hardware and typing in your first line of code. Well-defined diagrams will not only provide a vivid illustration of your application flow but will also serve as a guide to organizing the code generation logic.

9

Using AI Tools to Draw Application Flow Diagrams

In the last chapter, we explored 10 beginner-friendly IoT projects, complete with ChatGPT prompt examples. Hopefully, you're eager to choose one and embark on your innovative development journey.

This chapter suggests beginning your first practice project by designing a flow diagram for your application before you start coding. This initial step is crucial. We'll guide you on how to use AI tools to create detailed diagrams that illustrate how the application works locally, connects to the internet, and accesses the cloud. These diagrams will help you understand data interaction, error handling, and the integration of different components of your IoT project. By the end of this chapter, you'll be able to translate the narrative of your project's application into a comprehensive diagram with the help of AI tools.

We will be covering the following topics:

- Using diagrams for a better application journey
- Processing data locally
- Establishing an internet connection
- Sending sensor data to Cloud
- Data processing on the cloud

Using diagrams for a better application journey

A comprehensive application flow diagram is beneficial before starting to code. A detailed diagram can clarify the process of local data handling, the interaction between sensor data and local electronics, error handling, the power consumption mechanism, and the conditions for activating the wireless connectivity setup. Once the data reaches the cloud, a diagram should be created to outline the flow for data storage, forwarding, processing, analysis, visualization, and customer notifications.

Compose a thorough narrative that precisely describes your *story* about data flow. This story will serve as a guide, providing clear instructions to ensure your code aligns with your specific needs and objectives.

The raw data generated from sensors typically forms small packets ranging from a few bits to bytes. It seems like a simple, easy, and straightforward journey where the MCU captures this data and sends it to its wireless interface to access the internet, which ultimately reaches the cloud. However, in actual deployment, this journey is rarely peaceful and joyful, always encountering many abnormal situations and uncontrollable challenges.

Therefore, before writing your first line of code, it's crucial to create a diagram that illustrates how data will be processed at each stage. This not only gives you a clear understanding of how the application works but also serves as a guiding baseline for the subsequent coding process with ChatGPT.

There are numerous AI tools available that can assist you in creating such diagrams. For a simple text block format diagram, you can interact with ChatGPT-4o (free). Alternatively, you can use GPTs such as *Diagram: Show Me* within ChatGPT 4 (this requires a subscription). For a more professional diagram, consider using dedicated AI-driven websites such as https://www.mermaidchart.com/. In the following sections, we'll practice generating diagrams for typical scenarios encountered in the IoT development journey, using the AI chat feature on www.mermaidchart.com.

Processing data locally

Sensor data is first collected locally by the MCU. In most cases, an MCU processes sensor data locally under the following situations:

- **Normal situation**: The MCU successfully fetches data from the sensor, or the sensor triggers an event as expected.

- **Abnormal situation**: The MCU fails to fetch data from the sensor, or the sensor fails to trigger an event. In this case, the most likely causes are sensor initialization failure or hardware malfunction.

To enhance the user experience and quickly indicate the service status, consider using a color-changing LED and a buzzer beep on your IoT device to convey meaningful messages to the user.

With this in mind, let us use the AI chat at https://www.mermaidchart.com/ to generate a flow.

The following screenshot shows the process to prompt the AI chat at mermaidchart.com to create a diagram following your instructions.

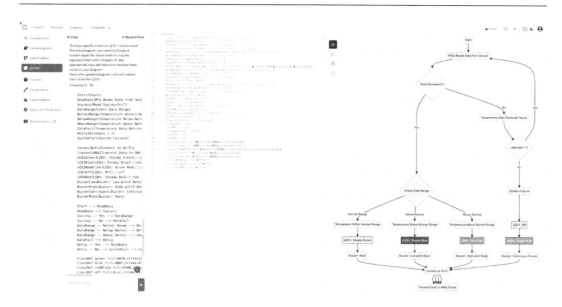

Figure 9.1 – An overview of how diagrams are generated by mermaidchart.com

You can use the following prompts as examples of instructions to insert in the **AI Chat** window:

```
Task: Imagine you're planning to develop a temperature sensor
end-device for monitoring conditions in customer warehouses. This
device includes a temperature sensor, an MCU, four distinct LEDs
for various status indicators (temperature range, Wi-Fi access, AWS
Cloud access, and system information), a buzzer, and an embedded
Wi-Fi module.
```

```
Action: Create a diagram illustrating the sequence of local data
interactions:
```

1. When MCU successfully reads the sensor's data, it will perform
    the following operation

    -   If the data is within Normal Range: LED1(Temperature Indicator)
        displays a steady green; mute the buzzer

    -   If the data is below Normal Range: LED1 changes to a steady
        blue; sounds the buzzer a low-pitch beep;

    -   If the data is above Normal Range: LED1 blinks red; sounds
        the buzzer a high-pitch beep;

2. When MCU fails to read the sensor's data, it will perform the following operation

    *   If data retrieval fails, the MCU will attempt 3 times before entering a failure state, LED#1 changes to off, LED4 (System Error Indicator) switches to a steady red, and sound the buzzer continuously

Goal: Ensure the diagram is clear, accurately reflecting the device's data interaction flow. It should visually represent sensor data processing, LED responses, Wi-Fi and AWS Cloud connectivity, and buzzer alerts based on different temperature readings.

The generated diagram is shown as follows.

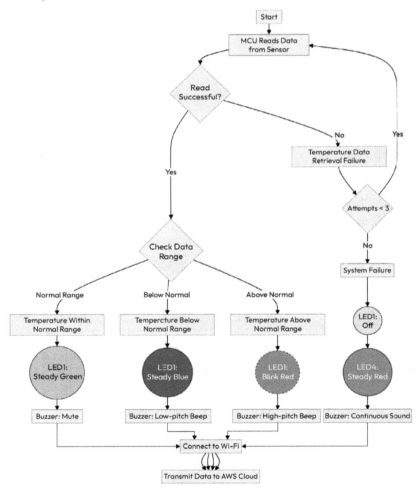

Figure 9.2 – Data local processing flow

In this section, we designed a data local processing flow with the collected value, the color changes of the LED, and the buzzer beeping on/off. As shown in *Figure 9.1*, we assumed to use the local Wi-Fi network to report the data to the AWS cloud. However, in an actual deployment, the internet access by a local Wi-Fi network still has an abnormal situation, such as a firewall behind the Wi-Fi network may block the device to access internet.

Establishing an internet connection

After processing local data, establishing internet connectivity is necessary to report the data to the cloud. To conserve power, the IoT device typically doesn't always need to maintain online connectivity. Instead, it should establish an internet connection only when it has data to transmit.

Assuming the use of a home Wi-Fi network to access the internet, you may encounter two basic situations:

- **Normal situation**: The MCU successfully connects to your home Wi-Fi router and obtains a valid IP address

- **Abnormal situation**: The MCU fails to connect to your home Wi-Fi router (not assigned a valid IP address) due to reasons such as weak signal strength or incorrect SSID and password

Continuing with `https://mermaidchart.com`, a flow diagram can be generated, as shown in the following chart. The following prompt was used to generate the diagram:

```
Task: Imagine you're planning to develop a temperature sensor
end-device for monitoring conditions in customer warehouses. This
device includes a temperature sensor, an MCU, four distinct LEDs
for various status indicators (temperature range, Wi-Fi access, AWS
Cloud access, and system information), a buzzer, and an embedded
Wi-Fi module. You are now focusing on the data connectivity sequence
by the following aspects:

Action: Create a diagram illustrating the sequence of Wi-Fi
connectivity. The MCU initiates connection to the home Wi-Fi router
after collecting data from sensor.

    • Normal Situation: If IP address is assigned by home Wi-Fi
      router, LED2 shows steady green.

    • Error Handling : If the MCU fails to connect to the Wi-Fi
      router, it will make three additional attempts. Upon failure,
      LED2 will illuminate steady red to indicate connectivity issues.

Goal: Construct a diagram that clearly delineates the device's
operational flow, including LED signal interpretation and Wi-Fi
connection. The diagram should effectively convey how the device
responds to various data conditions, manages Wi-Fi connectivity, and
indicates system status through LED color settings.
```

The generated diagram is shown as follows:

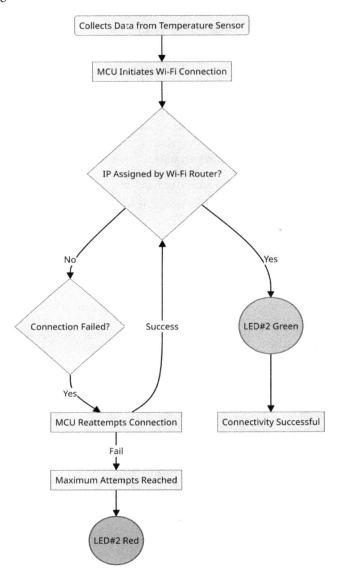

Figure 9.3 – Wi-Fi access flow

In this section, we designed a flow to access a home Wi-Fi network considering both normal and abnormal situations. Now, given successful access to the internet, we expect the sensor data to arrive at the cloud. However, there may be exceptions to this process. For instance, the device can be rejected by the cloud due to improper authentication credentials, we need to be aware of such scenarios.

Sending sensor data to Cloud

Once wireless connectivity is established and an IP address is granted, your device will be ready to transmit data to the cloud. However, there may still be two situations to consider at this step:

- **Normal situation**: The MCU successfully connects to the cloud

- **Abnormal situation**: The MCU fails to connect to the cloud due to some reasons, such as incorrect registration credentials

Using `https://mermaidchart.com`, you can have a data flow diagram as shown in the following chart. The following prompt was used to generate the diagram:

```
Task: Imagine you're planning to develop a temperature sensor
end-device for monitoring conditions in customer warehouses. This
device includes a temperature sensor, an MCU, four distinct LEDs
for various status indicators (temperature range, Wi-Fi access, AWS
Cloud access, and system information), a buzzer, and an embedded
Wi-Fi module. You are now focusing on the data arrivals at the Cloud,
considering the following aspects:

Action: Create a diagram illustrating the sequence of data arrivals
at the Cloud, i.e., AWS IoT Core.

  1. Cloud Connectivity:

    · The MCU Initiate and establish a secure MQTT communication
      between the MCU and AWS IoT Core for data transmission.

  2. Data Handling:

    · Normal Situation: The MCU receives acknowledge from AWS IoT
      Core and set LED3 to steady green.

    · Error Handling : In scenarios where the MCU does not receive
      acknowledgment from AWS IoT Core, it attempts to resend the
      data three more times. If all attempts fail, LED3 is activated
      to a steady red, signaling an issue with cloud connectivity.

Goal: This diagram should serve as a visual guide, clearly conveying
the steps involved in data transmission to AWS IoT Core, including
the mechanisms for handling communication success or failure.
```

The generated diagram is shown as follows.

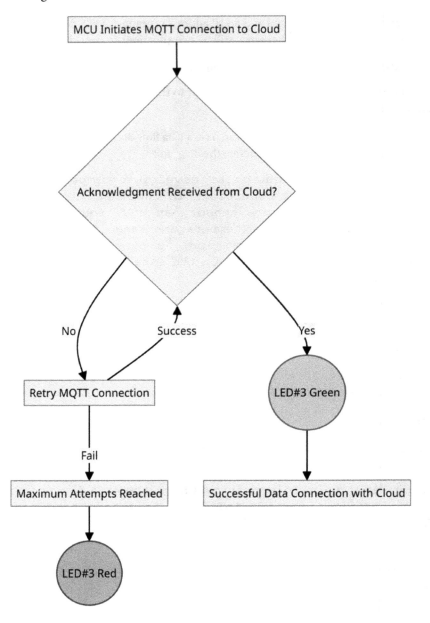

Figure 9.4 – Cloud access flow

Data processing on the cloud

Once the device establishes communication with the cloud and sends the data, the data starts its journey within the cloud. You can still use https://mermaidchart.com to create a data process flow diagram outlining how the data is processed:

Task: Imagine you're planning to develop a temperature sensor end-device for monitoring conditions in customer warehouses. This device includes a temperature sensor, an MCU, four distinct LEDs for various status indicators (temperature range, Wi-Fi access, AWS Cloud access, and system information), a buzzer, and an embedded Wi-Fi module. The current phase of development focuses on the data handling and processing within the AWS Cloud environment.

Action: Design a comprehensive diagram that illustrates the sequence of data handling and processing among various AWS Cloud services.

1. Data Ingestion:

   - The sensor data is published to the AWS IoT Core through MQTT protocol

2. Data Processing and Analysis:

   - The AWS IoT Core forwards the abnormal sensor data to AWS Lambda for processing and analysis.

   - When determines abnormal situation, AWS Lambda triggers an alert notification to AWS SNS.

   - AWS SNS sends email or message to notify customer

3. Data Storage, enrich, and Query:

   - AWS IoT Core routes all the sensor data to AWS IoT Analytics

4. Data Visualization:

   - AWS QuickSight query sensor data from AWS IoT Analytics to generate customer application dashboard.

Goal: Construct a diagram that meticulously details the data processing flow across different AWS services. This diagram should not only depict the operational workflow but also clarify how data moves from the sensor to the cloud, undergoes processing and analysis, is stored, visualized, and triggers alert notifications. Emphasize the integration and interaction between AWS IoT Core, AWS Lambda, AWS

IoT Analytics, AWS SNS and AWS QuickSight to provide a clear and comprehensive view of the system's architecture and data lifecycle.

The generated diagram is shown as follows.

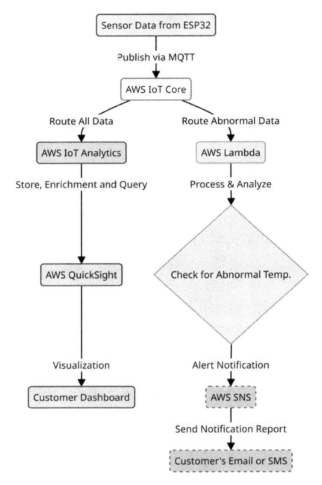

Figure 9.5 – Data processing on the cloud

After generating a diagram that accurately represents your application's flow, you'll have a comprehensive understanding of its functionality, from the sensor to the cloud. This overview not only provides a visual representation of your application but also helps identify potential areas for improvement or issues. Equipped with this visual aid and a clear understanding of your application's workflow, you're ready to begin your innovation journey. The next crucial step is setting up your development environment before creating your first project.

Summary

In this chapter, we started your journey toward IoT innovation. With the help of AI-driven tools such as `https://mermaidchart.com`, you can visualize clear application flows. This aids in grasping the logic of service interactions, including how to manage abnormal situations. Not only does this diagram enhance your understanding from a system design perspective, but it also serves as a streamlined guide for ChatGPT to comprehend your flow and generate code accordingly.

As we move on to the next chapter, we will be focusing on setting up a development environment. This will be accomplished by utilizing the PlatformIO IDE within Visual Studio Code. This setup is crucial as it facilitates the process of compiling and uploading the software code generated by ChatGPT. More specifically, in the subsequent example project, you will gain hands-on experience with this environment, learning how to effectively use it to compile and upload the software code. This practical exercise will not only consolidate your understanding of the development process but also equip you with the necessary skills to handle future IoT projects.

Part 3: Practicing an End-to-End Project

This part provides a thorough guide on establishing a development environment using Visual Studio Code and the PlatformIO IDE, focusing on software installation and setup for IoT projects. It introduces ChatGPT-assisted C++ programming for an ESP32 microcontroller, covering code writing, compilation, and debugging. Additionally, the guide details using ChatGPT for connecting the ESP32 to Wi-Fi and integrating it with AWS IoT Core via MQTT/TLS. You will learn how to transmit sensor data to AWS IoT Core, utilize ChatGPT-assisted Python coding for data processing with AWS Lambda, and manage data storage and queries using AWS IoT Analytics. Finally, it teaches creating interactive data visualization dashboards on ThingsBoard Cloud, equipping you with the ability to confidently manage your own comprehensive IoT projects from start to finish.

This part contains the following chapters:

- *Chapter 10, Setting Up the Development Environment for Your First Project*
- *Chapter 11, Programming Your First Code on ESP32*
- *Chapter 12, Establishing Wi-Fi Connectivity*
- *Chapter 13, Connecting the ESP32 to AWS IoT Core*
- *Chapter 14, Publishing Sensor Data to AWS IoT Core*
- *Chapter 15, Processing, Storing, and Querying Sensor Data on AWS Cloud*
- *Chapter 16, Creating a Data Visualization Dashboard on ThingsBoard*

10

Setting Up the Development Environment for Your First Project

In *Chapter 8*, we covered 10 beginner-friendly IoT projects with ChatGPT prompt examples. You might be excited to select one and start your innovative development journey. In *Chapter 9*, we went through the application diagram generation approach by using AI-driven tools.

In this chapter, we aim to empower you with the practical skills required to translate your innovative IoT concepts into reality. Learning how to set up a development environment using VS Code, PlatformIO IDE, and other coding extensions will provide you with the tools necessary to compile and upload software code effectively. Furthermore, creating your first project in PlatformIO will offer hands-on experience, deepening your understanding of the development process. These skills will not only facilitate your current learning journey, but will also equip you for future IoT projects, enhancing your ability to bring your creative visions to life.

In this chapter, we'll cover the following topics:

- Installing **Visual Studio Code (VS Code)**
- Setting up PlatformIO IDE
- Installing other coding assistance extensions
- Creating your first project under PlatformIO

Technical requirements

This chapter will show you how to install VS Code, PlatformIO, and other coding assistance extensions on a MacBook. To get optimal results, make sure you have your system equipped with an ARM CPU (i.e., Apple M1) running macOS Sonoma, version 14.3.1.

Installing Visual Studio Code (VS Code)

Visual Studio Code, commonly referred to as **VS Code**, is a lightweight yet powerful source code editor developed by Microsoft. It offers built-in support for JavaScript, TypeScript, and Node.js, along with a rich ecosystem of extensions for other languages such as C++, C#, Python, PHP, and more. Furthermore, it provides features such as debugging, syntax highlighting, intelligent code completion, snippets, code refactoring, and embedded Git.

In this chapter, we are going to instruct the VS Code installation on macOS. You may find the installation guidance on Windows and Linux from the internet.

To kick off the setup for our project, let's look at the following process and install VS Code on macOS:

1.   First and foremost, you need to download the VS Code software from `https://code.visualstudio.com/` by clicking on **Download Mac Universal**.

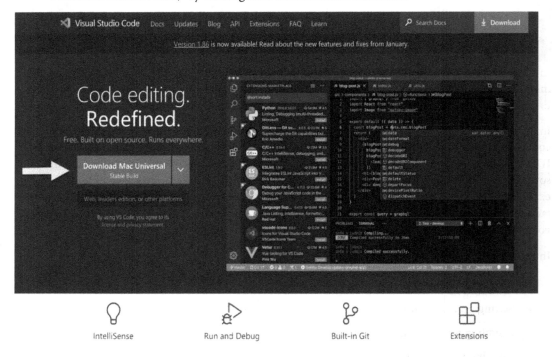

Figure 10.1 – Downloading VS Code

2.   Click on the downloaded **Visual Studio Code** software in your system's **Downloads** directory.

Figure 10.2 – Finding VS Code in Downloads

3. You will now see the **Welcome** page. From here, you can choose a theme according to your preference.

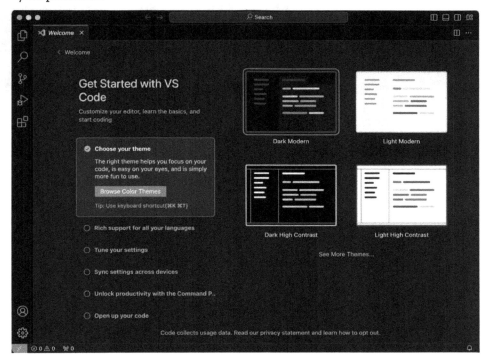

Figure 10.3 – VS Code Welcome page

4. On the left sidebar, find the *Extensions* icon (arrowed in *Figure 10.4*). We will install PlatformIO and other coding assistance tools from this *Extension* section.

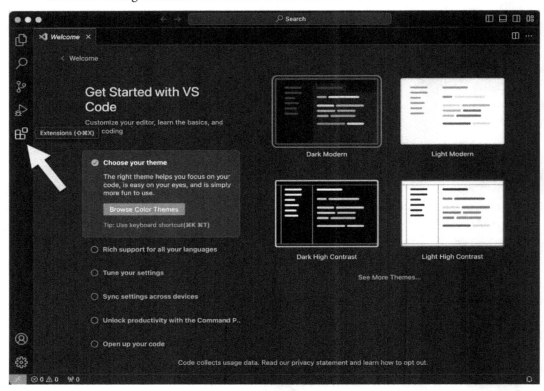

Figure 10.4 – Extensions in the VS Code Welcome page

You have now completed the initial installation of VS Code. The next step is to install the PlatformIO IDE extension on VS Code.

Setting up PlatformIO IDE

An open source **Integrated Development Environment** (**IDE**), **PlatformIO** works cross-platform and cross-architecture. It supports over 30 embedded platforms, features a multi-platform build system, includes numerous libraries, and supports over 800 open source hardware boards. It also serves as a powerful extension for coding in VS Code. In the following steps, we will learn how to install PlatformIO:

1. Continuing from *step 4* of the previous section, click on **Extensions**. You'll find the search window at the upper left, as visible in *Figure 10.5*.

Figure 10.5 – Extension search window

2. Type `Platformio` into the search window to locate it.

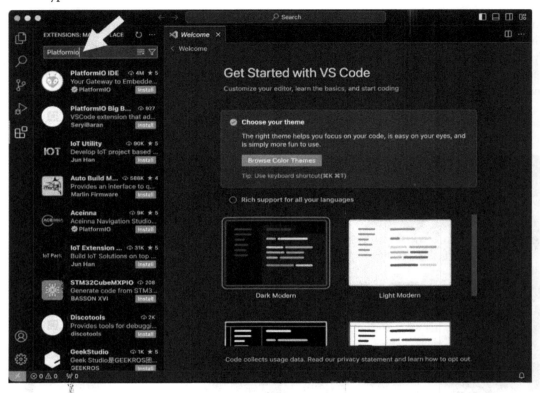

Figure 10.6 – Searching "Platformio" in search window

3. Click **Install** under **PlatformIO IDE** to install this extension.

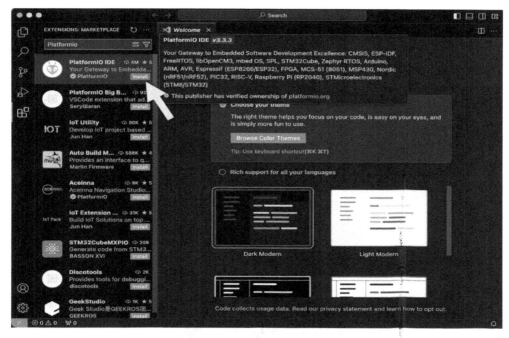

Figure 10.7 – Select and install "PlatformIO IDE"

4. After the installation, you will see the **Welcome** page of **PlatformIO** as follows:

Figure 10.8 – PlatformIO IDE welcome page

5.   To initiate PlatformIO for the first time, click on the icon resembling an *ant's head*. This action will start the PlatformIO initialization process. This process includes the installation of PlatformIO Core and other required software packages such as Python and *clang*. Please install them as required and note that this might take several minutes.

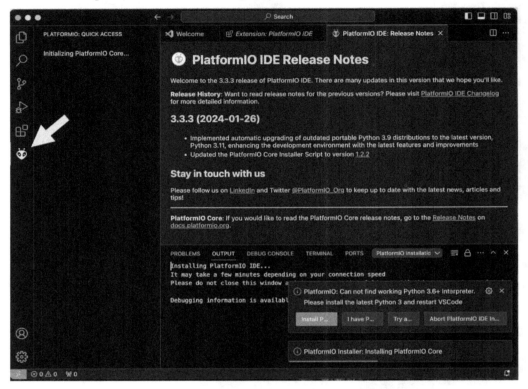

Figure 10.9 – First launch of PlatformIO after installation

6.   Allow *clang* to be installed by clicking **Install** if it hasn't been installed on your MacBook before.

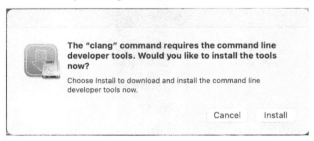

Figure 10.10 – Clang tools installation

7.   Continue the installation process on the next screen, as shown in *Figure 10.11*.

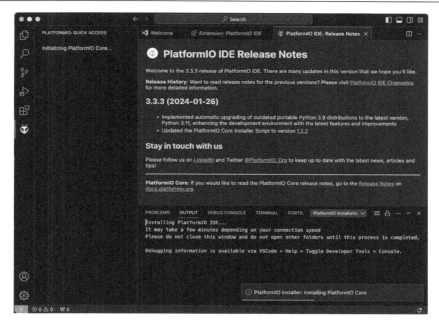

Figure 10.11 – PlatformIO Core initialization process

8.   Click **Reload Now** after the installation.

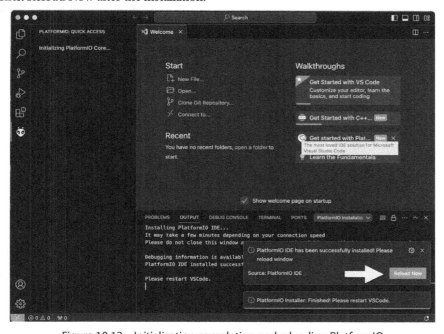

Figure 10.12 – Initialization completion and reloading PlatformIO

9. After reloading, you will see the PlatformIO interface as shown in *Figure 10.12*. Ccontinue clicking on **TERMINAL** and you will be directed to the PlatformIO terminal window.

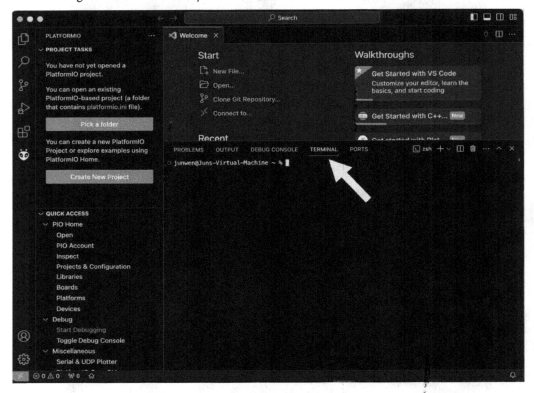

Figure 10.13 – Open the terminal window

10. Now we are going to install the open source software package manager. Linux systems such as Ubuntu and Debian use **Advanced Package Tool (APT)** to handle the installation and removal of free software package. In macOS, Homebrew is the most popular one. If **Homebrew** is not yet installed on your MacBook, the next step is to install it.

Homebrew is a popular open source package management system that simplifies the installation of software on Apple's macOS and Linux operating systems. It offers a convenient and streamlined process, eliminating the need for manual downloads and installations. Use your browser to navigate to `https://brew.sh/`, and then, as indicated by the arrow in the following figure, click the copy icon that will *copy* the homebrew installation link, `/bin/bash -c "$(curl -fsSL https://raw.githubusercontent.com/Homebrew/install/HEAD/install.sh)"`.

Figure 10.14 – Copy the Homebrew installation link

11. Switch to your VS Code window, paste the installation link in the **TERMINAL** command line, and press *Enter*.

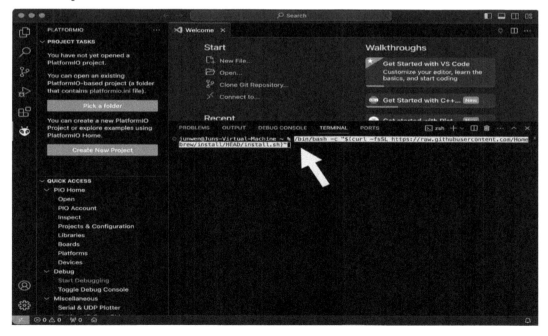

Figure 10.15 – Paste the Homebrew installation link in the Terminal window of PlatformIO

12. You may be requested to type in your root user password, which is the login password for your MacBook.

Figure 10.16 – Type in your root user password

13. The installation process of Homebrew will start. Then, press *RETURN* or *ENTER* to continue when required.

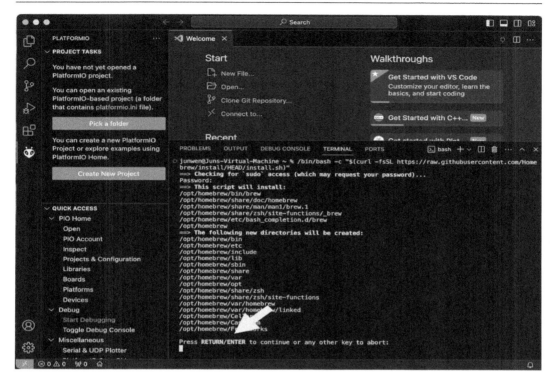

Figure 10.17 – Start Homebrew installation on PlatformIO

14. After the installation process of Homebrew is complete, follow the indication to add Homebrew to your PATH:

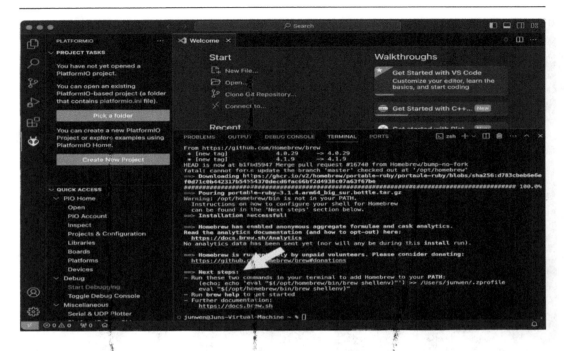

Figure 10.18 – Add Homebrew to the PATH of your MacOS

15.   Copy, paste, and run the first command.

Figure 10.19 – Execute the first command

16. Then copy, paste, and run the second command.

Figure 10.20 – Execute the second command

17. In the Terminal window, install PlatformIO CLI (command line) with `brew install platformio`.

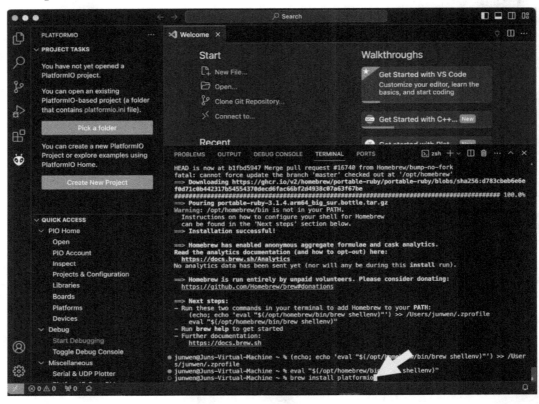

Figure 10.21 – Use Homebrew to install PlatformIO CLI

18. Upon completion of the PlatformIO CLI installation, restart your VS Code to use the PlatformIO command line in the Terminal window.

Figure 10.22 – Install the PlatformIO source through the CLI

19. After restarting, install the latest stable PlatformIO packages in the Terminal window via the CLI using `pio platform install espressif32`.

Figure 10.23 – Install PlatformIO latest package

20. The installation process should now be running as seen in the following figure.

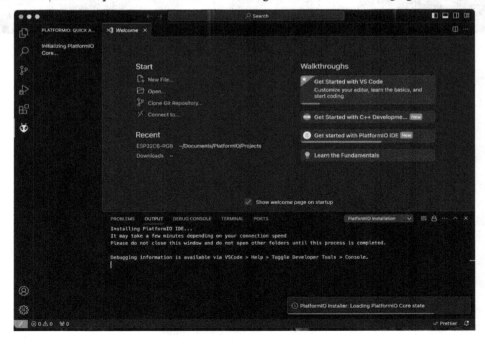

Figure 10.24 – PlatformIO package installation process

Once the installation process is complete, you need to restart VS Code. Congratulations! You have successfully set up the development environment and are ready to start your first project!

After installation, you can check your PlatformIO system information with the `pio system info` command in the Terminal window.

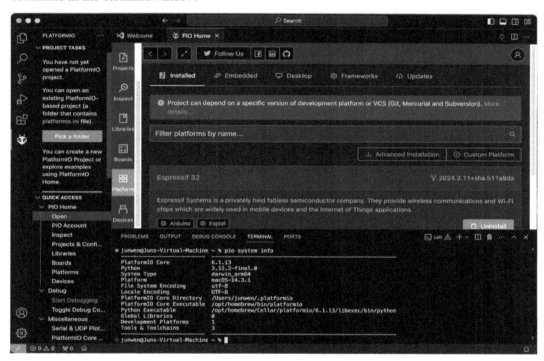

Figure 10.25 – Check the PlatformIO package information

We have now successfully completed the installation of the PlatformIO IDE extension on Visual Studio Code. This includes its CLI and the most recent package. This setup forms the foundational structure for our coding environment.

To optimize our development workflow and experience, it is essential to consider installing additional extensions. These extensions can significantly enhance your coding, troubleshooting, and testing processes.

Installing other coding assistance extensions

There are many coding assistance tools available as extensions in VS Code, such as Prettier, Indent Rainbow, and Better Comments. They can help make your code easier to read and debug, and enhance your coding experience:

- **Prettier**: A popular code formatter, it supports many languages and integrates well with VS Code. Prettier automatically formats your code according to a set of predefined style guidelines. This helps to keep your code looking clean and consistent, making it easier to read and maintain. You can configure it to format your code each time you save a file, or you can run it manually. This tool is especially useful in team projects to ensure that everyone adheres to the same coding style, thereby reducing discrepancies and improving collaboration.

- **Indent Rainbow**: This is a visual tool that makes indentation more readable by coloring indent levels in a gradient-like manner. Each indentation level is given a distinct color, which helps in distinguishing the scopes and blocks of code at a glance. This can be particularly helpful in languages where indentation plays a crucial role, such as Python, or in any complex nested code structure, making it easier to follow the logical flow of the code. The colors and the number of spaces considered as an indentation level are customizable, allowing users to tailor the appearance to their liking and coding standards.

- **Better Comments**: This enhances the readability and functionality of comments within your code. This extension allows you to categorize and color-code your comments, making them more noticeable and organized. For instance, you can differentiate comments that are informational, questions, TODOs, highlights, or alerts, each with different colors or formats.

Creating your first project under PlatformIO

After setting up the development environment, let's walk through the process of creating your first project in PlatformIO:

1. Click the PlatformIO icon in the left-hand bar.

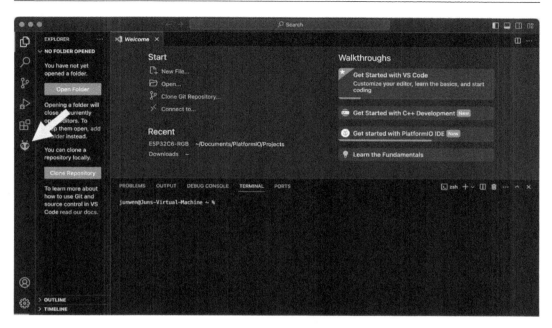

Figure 10.26 – Launch PlatformIO IDE in VS Code

2. Click **Open** under **PIO Home**.

Figure 10.27 – PlatformIO welcome page

3.   Click **New Project** under **Quick Access**.

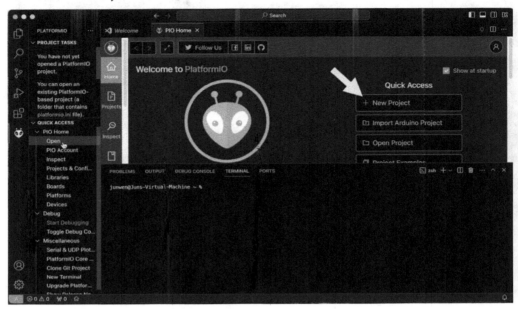

Figure 10.28 – Create a new project

4.   You will see **Project Wizard** as shown in the following figure. Give the project a name in the **Name** field.

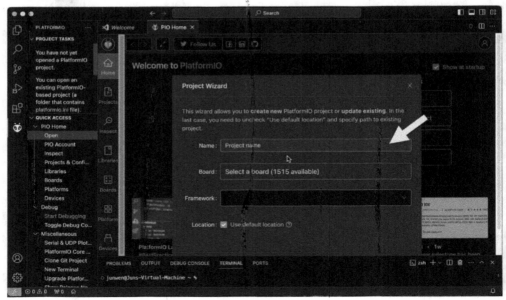

Figure 10.29 – Give the project a name

5.   Click the **Board** dropdown to find your ESP32 board model.

Figure 10.30 – Find the hardware type

For example, if you want to use ESP32-C3, you can type in esp32-c3, and you will find your expected board model. Click the model name to select it.

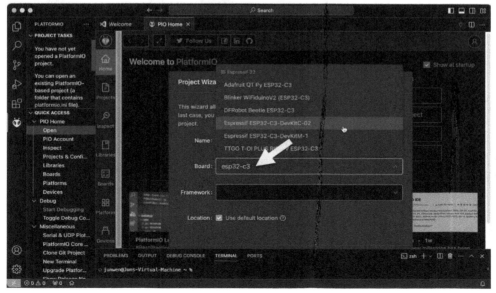

Figure 10.31 – Select the hardware you are going to use

6.   In the **Framework** dropdown, make sure to select **Arduino Framework**, and click **Finish**.

Figure 10.32 – Select "Arduino Framework" for the Framework dropdown

7.   Then PlatformIO starts to configure your project automatically as shown in the following figure.

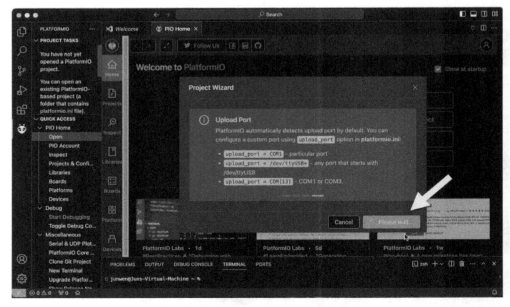

Figure 10.33 – New project being created

8. Click **Yes, I trust the authors** to continue.

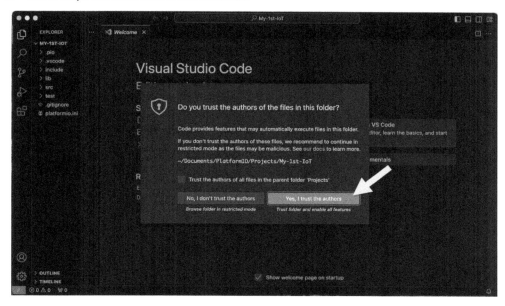

Figure 10.34 – Select "Yes" to trust the authors

9. Now your first project has been created correctly as seen in the following figure. Click **src** in the left-hand bar and you will see **main.cpp**, which is the place where you will program your C++ code snippets.

Figure 10.35 – Browse the main.cpp template

By now, all necessary development environment settings should be correctly configured. You should be able to successfully create your first IoT project in PlatformIO IDE.

Summary

In this chapter, you have set up your development environment and are ready to create your first IoT project. Now that you have configured VS Code and PlatformIO IDE, you have the tools and knowledge you need to embark on this innovative journey. In the upcoming chapter, we will guide you through the practical implementation of your first project, helping you to apply what you've learned effectively.

11

Programming Your First Code on ESP32

Having completed the setup of the development environment, VS Code, and the PlatformIO IDE extension in *Chapter 10*, along with your first project, it's time to start a hands-on practical case. Beginning with this chapter, we will design a temperature and humidity monitoring application. We will use an ESP32-C3 and a DHT11 sensor to construct a hardware prototype in this chapter. In the subsequent chapters, we will establish Wi-Fi connectivity, send data and process it on AWS, and create a visualization dashboard on ThingsBoard Cloud.

By the end of this chapter, you will be able to harness ChatGPT to complete an operational hardware prototype and collect the sensor data locally.

This chapter will cover the following topics:

- Designing the application's local logic
- Creating a flow diagram with ChatGPT
- Building a device hardware prototype
- Instructing ChatGPT to generate C++ code
- Code examples
- Using PlatformIO to program code on the ESP32

Designing the application's local logic

This section will teach you how to use your logical thinking to meticulously design the application flow. This process includes planning and strategizing for regular procedures, irregular scenarios, and fault situations. Think about how the application should respond, what visual and audio indicators should be displayed on the device, and which functions should be activated.

Imagine a scenario where we create an IoT project to monitor the temperature and humidity inside a warehouse. For this, we need to develop a sensor device with the following capabilities:

- Periodically measure the temperature and humidity data
- Report sensor data to the cloud
- Alert the customer if the sensor data is outside the normal range

In this section, our focus will be on the application logic on the device side. For an optimal user experience, we want the sensor device to not only periodically read the data but also provide visual and audible indicators via an LED and buzzer for warehouse workers. With these considerations in mind, let's design the local logic of the application, as follows:

1. The ESP32 retrieves temperature and humidity data from the DHT11 sensor periodically.
2. The ESP32 checks the retrieved data against a preset normal range.
3. Normal condition – If the retrieved data falls within the normal range, a green LED remains solid on without any alert beep.
4. Abnormal condition – If the retrieved data falls below the normal range, a blinking blue LED and a beep signal this. If the data exceeds the normal range, a blinking red LED and a beep indicate this.
5. Fault handling – If the DHT11 fails to provide data, a system LED turns on red, and a continuous sound indicates a failure state. The ESP32-C3 tries to read the data three times. If all attempts fail, it initiates a reboot.

This structured logic design ensures that the device not only functions efficiently but also communicates effectively with the warehouse staff, ensuring prompt and appropriate responses to environmental changes. With this local logic, you can use an AI tool as mentioned in *Chapter 10* to create a flow diagram.

Creating a flow diagram with ChatGPT

In this section, we'll create a comprehensive and visually engaging diagram to effectively reflect your application's flow. This process aids in conceptualizing your application's structure and serves as a reference for building code in the next step.

Using the logic flow designed in the last section, and creating a prompt text in the **AI Chat** window of `https://mermaidchart.com`, you may see a service flow diagram generated like the following. Please note that this diagram was *not* mentioned in *Chapter 9*.

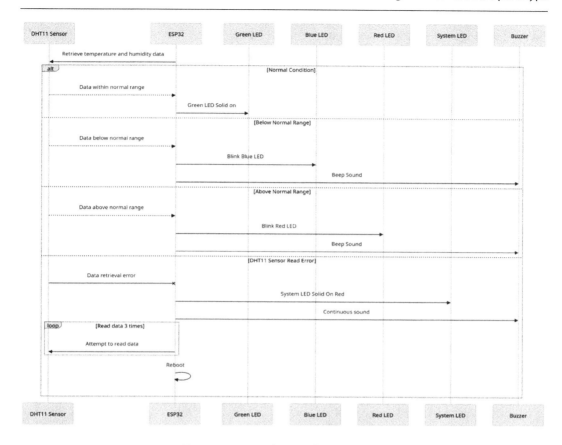

Figure 11.1 – Local service logic diagram

A clear and straightforward diagram will certainly help you establish a well-rounded system design, thoroughly considering both normal cases and abnormal contingencies. You can use this diagram as a guideline to craft your hardware prototype and instruct ChatGPT to generate code accordingly. Now, let's proceed to the section on creating a hardware prototype.

Building a device hardware prototype

To ensure that the Arduino-compatible temperature and humidity sensor, DHT11, operates correctly with the MCU, ESP32, for data collection, you must understand the correct wire connections between them. Additionally, you should correctly pair the wires between the ESP32, a buzzer for alerts, and an LED to indicate the data range. According to the diagram we created in the last section, to craft the device hardware prototype, we need the following elements:

- An MCU – ESP32-C3
- A temperature and humidity sensor – DHT11

- A data range indicator LED with red, green, and blue colors

- A system LED on the ESP32-C3

- A piezo buzzer

Let's understand each of these in greater detail:

- **ESP32-C3**: In *Chapter 3*, we introduced the ESP32-C3, including its specifications and interfaces, such as GPIO, SPI, and I2C. The ESP32-C3 (EVB model: esp32-c3-devkitc-02) module's PIN layout is shown in the following table.

PIN Order	PIN Name	GPIO	ADC	I2C	SPI	Other Functions
1	GND					
2	IO00	GPIO0	ADC_0			
3	IO01	GPIO1	ADC_1			
4	IO12	GPIO12			SPI_HD	LED D4 control
5	IO18	GPIO18				USB_D-
6	IO19	GPIO19				USB_D+
7	GND					
8	UART0_RX	GPIO20				
9	UART0_TX	GPIO21				
10	IO13	GPIO13				
11	NC.					
12	RST					RTC
13	3V3					
14	GND					
15	PWB					
16	5V0					
17	GND					
18	3V3					
19	IO02	GPIO2	ADC_2		SPI_CK	
20	IO03	GPIO3	ADC_3		SPI_MOSI	
21	IO10	GPIO10			SPI_MISO	

PIN Order	PIN Name	GPIO	ADC	I2C	SPI	Other Functions
22	IO06	GPIO6				
23	IO07	GPIO7			SPI_CS	
24	IO11	GPIO11			VDD_SPI	
25	GND					
26	3V3					
27	IO05	GPIO5	ADC_5	I2C_SCL		
28	IO04	GPIO4	ADC_4	I2C_SDA		
29	IO08	GPIO8				
30	BOOT	GPIO9				
31	5V0					
32	GND					

Table 11.1 – ESP32-C3 pinout map

Please note that the ESP32-C3 features two built-in LEDs:

- LED D4, which is controlled by GPIO12
- LED D5, which is controlled by GPIO13

In this chapter, we will use LED D5 as the system LED. In *Chapter 12*, we will configure LED D4 as the indicator for internet access status indicator.

- **DHT11**: The DHT11 temperature and humidity sensor was introduced in *Chapter 8*, equipped with GND, VCC, and an output data port.
- **Piezo buzzer**: The piezo buzzer we use here is of the active type, requiring both GND and VCC inputs.
- **RGB LED**: The RGB LED we use here is a color-changing LED. It is equipped with GND, VCC, red, blue, and green pins.
- **Wires connection**: The following schematic diagram will guide you in wiring the ESP32-C3 module with a DHT11 sensor, a piezo buzzer, and an RGB LED.
- **Power supply**: The power supply to this board is through the USB-C port when you connect it through a USB-C cable from your MacBook.

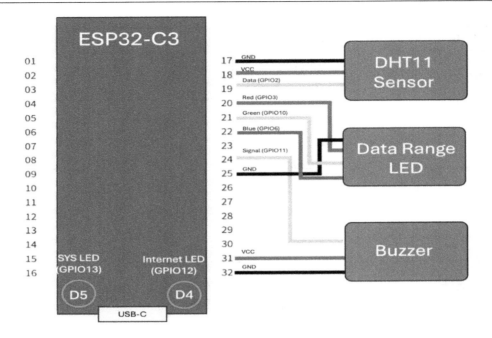

Figure 11.2 – Wires connection schematic diagram

By now, you should be able to correctly wire the DHT11 sensor, the data range indicator LED, and the buzzer to the ESP32 according to the schematic diagram in *Figure 11.2*. With the hardware prototype built, you are now ready to start coding on the ESP32.

Instructing ChatGPT to generate C++ code

In this section, we will direct ChatGPT to produce C++ code on ESP32, using the well-designed application logic diagram and the hardware pin wire connections as effective prompts.

After you have carefully reviewed the diagram, understood the application logic, and familiarized yourself with the ChatGPT prompts outlined in *Chapters 6* and *8*, you can create custom prompts. These prompts will instruct ChatGPT to generate code in accordance with your application logic.

The following is a prompt example to instruct ChatGPT to generate code according to your application logic:

```
Hi, ChatGPT,

Role:

You act as a senior software developer with expertise in embedded
development, specifically with IoT projects using the ESP32, Arduino-
compatible sensors, and AWS Cloud.

Task:

Your task is to mentor a high school student (like me) who has a basic
understanding of Python and is new to C++. I need you to develop a
comprehensive code snippet that fulfills the following objective,
detailed requirements, implementation guidance, and output format.

Objective:

Create an educational C++ code snippet on ESP32-C3 using the PlatformIO
IDE with Arduino framework and Espressif32 platform, following the
requirements below.

Requirements:

  1.  Pin Wire Connections:

      I.    Connect the DHT11 data pin to IO2 on the ESP32-C3.

      II.   Connect the Piezo Buzzer signal pin to IO11 on the ESP32-C3.

      III.  Connect the RGB LED to IO1 to control Red color, IO12 for
            Blue color, and IO0 for Green coloron the ESP32-C3.

      IV.   Use IO13 to control the ESP32-C3's built-in LED D5, designated
            as the System LED.

  2.  Data Retrieve Operation: Periodically read the DHT11 sensor's
      temperature and humidity data and print it out locally in both
      Celsius and Fahrenheit.

  3.  Normal Condition: The retrieved data falls within the pre-defined
      normal range.

  4.  Abnormal Condition: The retrieved data falls out of the pre-defined
      normal range.

  5.  Error Condition: Data retrieval fails, and the system will then
      attempt to read the data three times. If all three attempts
      fail, the system will initiate a reboot.
```

6. Visual and Sound Indication:

   I.   Normal Condition: RGB LED turns solid on and mute the buzzer.

   II. Abnormal Condition: If the data exceeds the normal range, the RGB LED will blink red, and the buzzer will beep in sync; If the data falls below the normal range, the RGB LED will blink blue, and the buzzer will beep in sync.

   III. Error Condition: Turns the system LED on solidly and triggers the buzzer to beep continuously.

Must Apply:

- Adopt C++ programming best practices.

- Use "constexpr" to declare hardware related variables.

- Uses the millis() function to manage timing without blocking other code execution

- Favor functions over classes.

- Implement PWM control for both the System LED and buzzer using ledcSetup and ledcAttachPin.

- Avoid using magic numbers when defining constant variables.

- Use a non-blocking approach to blink the LED independently.

- Include necessary dependency libraries.

- Provide line-by-line comments for clarity.

Output Format: Your code snippets output must comply with the following example format.

```
1.  // *********************************
2.  // Created by: ESP32 Coding Assistant
3.  // Creation Date: [Current Date]
4.  // *********************************
5.  // Code Explanation
6.  // *********************************
7.  // Code Purpose:
9.  // Requirement Summary:
11. // Hardware Connection:
13. // New Created Function/Class:
15. // Security Considerations:
17. // Testing and Validation Approach:
19. // *********************************
```

```
20. // Libraries Import
21. // *********************************
23. // *********************************
24. // Constants Declaration
25. // *********************************
27. // *********************************
28. // Variables Declaration
29. // *********************************
31. // *********************************
32. //  Declaration
33. // *********************************
35. // *********************************
36. // Setup Function
37. // *********************************
39. // *********************************
40. // Main loop Function
41. // *********************************
42. // *********************************
40. // Functions Definition
41. // *********************************
```

- Create a platformio.ini file that incorporates the required library dependencies and environment settings.

- Ensure the following information is included.

```
1. [env:esp32-c3-devkitc-02]
2. platform = espressif32
3. board = esp32-c3-devkitc-02
4. framework = arduino
5. monitor_filters = esp32_exception_decoder, colorize
6. monitor_speed = 115200
7. build_src_filter = +<../../src/>  +<./>
8. board_build.flash_mode = dio
9. build_flags =
10.     -DARDUINO_USB_MODE=1
11.     -DARDUINO_USB_CDC_ON_BOOT=1
12.     -w
13. lib_deps =
14.     adafruit/DHT sensor library@^1.4.6
15.     adafruit/Adafruit Unified Sensor@^1.1.14
```

By now, you have learned how to use these prompts to request ChatGPT to generate code on your ESP32. Next, let's look at some code examples.

Code examples

You can find the examples of `ChatGPT_Prompt`, `main.cpp` code and the `platformio.ini` file in `https://github.com/PacktPublishing/Accelerating-IoT-Development-with-ChatGPT/tree/main/Chapter_11`.

In the example of `main.cpp`, you can find the user-defined `setup()` and `loop()` functions, which essentially run within a FreeRTOS task. We mentioned FreeRTOS in the *MCUs* section of *Chapter 3*. ESP32 supports FreeRTOS by default when using the Arduino framework on PlatformIO. The Arduino ESP32 platform automatically includes a FreeRTOS stack and calls the APIs of FreeRTOS in the `setup()` and `loop()` functions.

In the `main.cpp` example code, the following seven functions are generated by ChatGPT. Please note that the function names might be different from the output of ChatGPT:

- `checkSensorReadings()`: To read the sensor data value
- `updateLEDs(bool red, bool green, bool blue)`: To update the RGB LED's color according to the data value
- `indicateNormalCondition()`: Indicate normal conditions when the data value falls in the range of a predefined threshold
- `indicateConditionBelowRange()`: Indicate condition below the range if the data value is beneath the predefined threshold
- `indicateConditionAboveRange()`: Indicate condition above the range if the data value exceeds the predefined threshold
- `indicateSensorError()`: Indicate sensor error if the data value is null
- `ledBlinking()`: Handle LED blinking and buzzer beeping

In addition, to control the LED and buzzer, as requested, ChatGPT creates the correct approach through **power width modulation (PWM)**. Compared to the traditional `digitalWrite()` method, PWM can effectively control an LED's brightness. This is preferable to using resistors to dim LEDs, as it is more energy-efficient and can achieve smoother transitions between brightness levels. Also, PWM can be used not just for the buzzer's volume control but also to generate different tones. By varying the frequency of the PWM signal, you can change the pitch of the sound emitted by the buzzer, which is useful in applications requiring different alert tones or musical notes.

By the end of the following section, you should be able to prompt ChatGPT to produce code by following your instructions and the application logic. In the next section, we will go through the process of compiling and uploading the code on ESP32 through PlatformIO.

Using PlatformIO to program code on the ESP32

In this section, we will build and upload code onto the ESP32 via the PlatformIO IDE. Then, we will observe the messages printed locally on our Macbooks, and verify whether the outcome aligns with our logic.

Here are the steps that are to be followed:

1. Launch the PlatformIO IDE in VS Code, go to the project you made in *Chapter 10*, look for main.cpp under src in the project folder, copy the code you got from the ChatGPT conversation window, and put the code in main.cpp, as shown here:

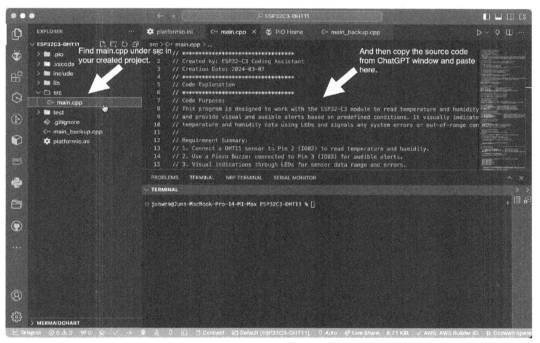

Figure 11.3 – Copying the main.cpp code from ChatGPT to PlatformIO

2. Locate the platformio.ini file in your project folder, then copy the platformio.ini contents from the ChatGPT conversation window and paste them in the platformio.ini file here.

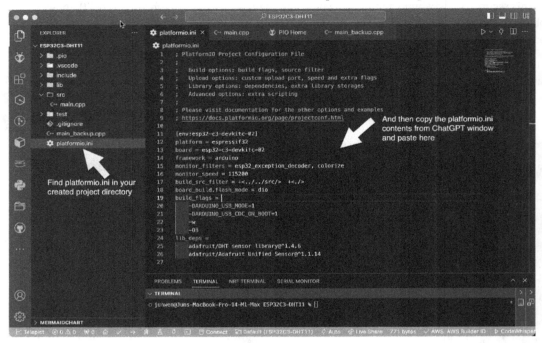

Figure 11.4 – Copying platformio.ini code from ChatGPT to PlatformIO

3. In the `platformio.ini` file example, there are two libraries under `lib_deps`: `adafruit/DHT sensor library@^1.4.6` and `adafruit/Adafruit Unified Sensor@^1.1.14`. You need to manually install them from the **Libraries** section of PlatformIO, as shown in the following screenshot.

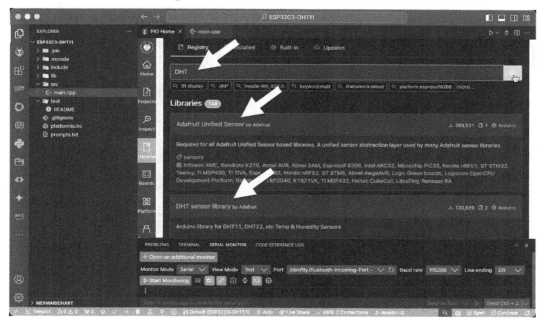

Figure 11.5 – Installing Adafruit Unified Sensor and DHT sensor library in PlatformIO

4.   Look for the *build* button, as shown in *Figure 11.6*, to compile the code produced from ChatGPT.

Figure 11.6 – The start of code building

5. In the **TERMINAL** window, you can watch the build process and outcome. If you see SUCCESS, as shown in *Figure 11.7*, it means your code is well compiled and ready to upload to the ESP32. If there are any errors reported, you can ask ChatGPT to help you correct them until the build succeeds.

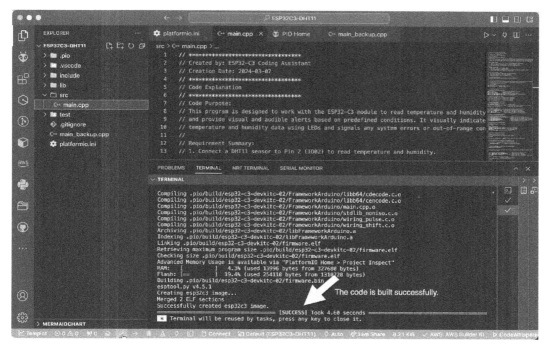

Figure 11.7 – Code building completion

6. After building the code successfully, make sure the console cable connection between your MacBook and the ESP32 is correct, following the instructions shown in *Figure 11.8* to check it, before uploading the code to the ESP32. Please note that the console cable is typically a Type-C to Type-C USB cable. If the console cable connection between your MacBook and the ESP32 board is correct, you can click *Auto* to check if the USB port is shown as follows.

Figure 11.8 – Checking the MacBook's USB port conntected to the ESP32

7.   Click **Upload** button to upload the compiled code to ESP32.

Figure 11.9 – Clicking "Upload" to upload the code to ESP32

8.   Watch the write process until it shows SUCCESS.

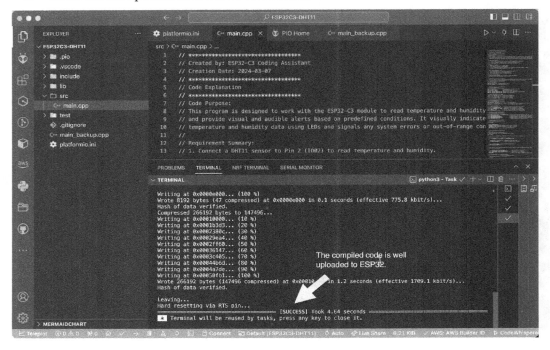

Figure 11.10 – Code uploading completion

7.   Click **Upload** button to upload the compiled code to ESP32.

Figure 11.9 – Clicking "Upload" to upload the code to ESP32

8.   Watch the write process until it shows SUCCESS.

Figure 11.10 – Code uploading completion

9. After the write process, the ESP32 will reboot and you can locate the **Serial Monitor** button to see the messages print out on the console port of your MacBook.

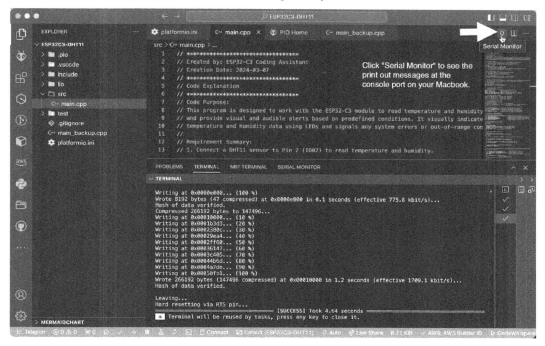

Figure 11.11 – Opening the local console window on Serial Monitor

10. When your code is right and matches what you expect, and you have successfully uploaded it to ESP32, you can see the messages printed out in the **TERMINAL** window.

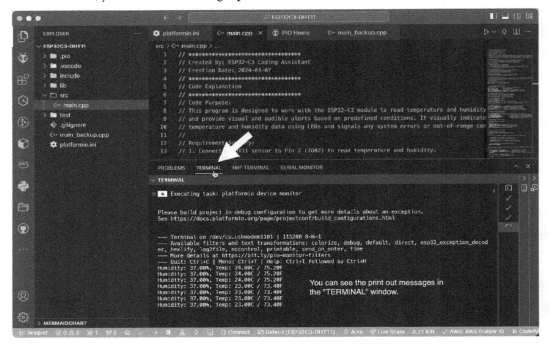

Figure 11.12 – Observing the output messages on the TERMINAL window

Now, you should be ready to compile and upload the code on the ESP32 to validate the outcome. As discussed in *Chapter 6*, ChatGPT may not provide 100% accurate code in the first round of conversation. You may need to continuously prompt ChatGPT to correct and improve its output through further conversation. Repeat the validation process several times until it fully aligns with your expectations.

Summary

In this chapter, you connected the ESP32-C3 to a DHT11 sensor and a piezo buzzer to successfully complete your first IoT project code. You should be able to validate the output on your local console port, observing the temperature and humidity data. The LED's color will change, and the buzzer will beep for abnormal or error situations following your application flow.

In the next chapter, we'll progress by connecting your ESP32-C3 to your home Wi-Fi network. This is a crucial step before transmitting DHT11 data to the AWS cloud.

12

Establishing Wi-Fi Connectivity

In the previous chapter, we successfully built the hardware prototype and programmed the first code to read data from the DHT11 sensor. This data can be observed from the local console port in the PlatformIO **TERMINAL** window. The next step is to establish Wi-Fi connectivity from the ESP32 to your home Wi-Fi router.

Connecting ESP32 to the internet is an imperative step to deliver sensor data to the cloud. In this chapter, we will continue to design the Wi-Fi access logic, draw a diagram using AI Chat at `https://mermaidchart.com`, and instruct ChatGPT to produce code to enable Wi-Fi access on ESP32.

This chapter will cover the following topics:

- Designing Wi-Fi access logic
- Creating the Wi-Fi access flow diagram
- Instructing ChatGPT to generate code
- Code examples
- Validating internet access on ESP32

In this chapter, we will store the Wi-Fi credentials, ping host address, and NTP server address in the `Platformio.ini` file, and then pass them to the main code. This is a highly adaptable methodology that has been designed to circumvent the need to directly store information within the main body of the code. Doing so makes future alterations and modifications easier, thereby improving the code's overall efficiency, functionality, security, and readability.

Designing Wi-Fi access logic

Imagine taking the hardware prototype we created in the last chapter to a warehouse. Here, you would need to connect the ESP32 to the local Wi-Fi network using the proper SSID and password. After receiving a valid IP address, you would ping a public host to verify internet access and then sync with an NTP server to get accurate clock timing. There are two ways to provide the SSID and password to your device:

- **Option 1**: Hardcode the SSID and password into the ESP32. This can be accomplished either by including them in the `main.cpp` code or by storing and passing them through the `Platformio.ini` file.

- **Option 2**: Use your mobile phone to assist ESP32 to connect to your local SSID and type in your password.

Option 1 is straightforward and simple but lacks security and flexibility. **Option 2** is flexible, secure, and easy to deploy at various sites, but it has a high level of complexity in the code design.

In this project, we'll use **Option 1** as an example to configure the SSID and password. Here's what the simplified Wi-Fi access logic looks like:

1. The ESP32 uses the SSID and password to access the local Wi-Fi **Access Point** (**AP**) or router.

2. The local Wi-Fi AP or router executes WPA2-PSK authentication with the ESP32.

   If authentication is successful, the local Wi-Fi AP or router assigns the ESP32 a valid local IP address. If the password is incorrect, the authentication fails.

3. Under normal conditions, once the ESP32 receives a valid IP address, it pings a public internet destination and synchronizes with an NTP server to get the current time.

4. Under abnormal conditions, should the ESP32 fail to obtain a valid IP address, the ESP32 is set to turn on the LED D4 as an indicator of internet connection failure, sound the buzzer, retry three times, and then reboot.

In this section, we designed logic for ESP32 to access a Wi-Fi network, taking into consideration normal conditions and abnormal contingencies. In the next section, we will use this logic to create a flow diagram with AI.

Creating the Wi-Fi access flow diagram

As with *Chapter 11*, we can continue to generate a diagram on `mermaidchart.com`, you may see a service flow diagram generated, as shown in the following example.

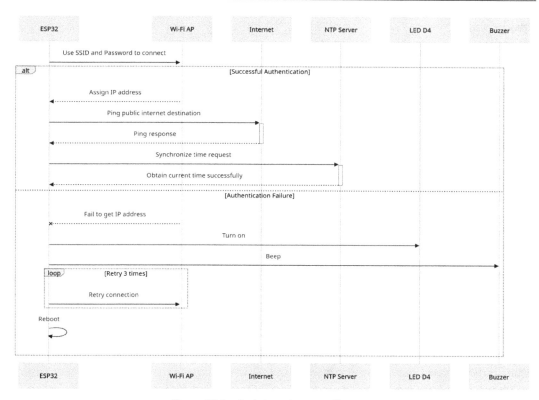

Figure 12.1 – An internet access diagram

In this diagram, we use the LED D4 on the ESP32 and the buzzer to indicate the status of internet access. They provide visual and auditory signals to easily show whether internet access is successful or not. This method is more convenient than observing printed messages in the **TERMINAL** window.

In this section, we created an internet access diagram with AI Chat on mermaidchart.com. By following this diagram, we can start to instruct ChatGPT to update the previous code in *Chapter 11* to support internet access through Wi-Fi.

Instructing ChatGPT to generate code

In this section, we will request ChatGPT to update the previous code generated in *Chapter 11* to implement the Wi-Fi access logic. You can instruct ChatGPT to work on the previous code and add internet access requirements:

```
Hi, ChatGPT

Please refer to the following code, maintain its current structure
and style, and support the following additional requirements.
```

Requirements:

- Wi-Fi access: Use the pre-provisioned SSID and password to access Wi-Fi. The SSID and password are stored in the Platformio.ini file.

- Normal condition: The ESP32 receives a valid IP address. It then pings a public internet destination and syncs with an NTP server to get the current time.

- Abnormal condition: If the ESP32 fails to obtain a valid IP address, the ESP32 LED D4 (IO12) will turn on as an indicator of Wi-Fi connection failure. The buzzer will beep, and the system will retry three times before rebooting.

Platformio.ini File:

- Create a new file that incorporates the ESP32Ping library.

Next, let's look at some examples of the code that ChatGPT can generate for us.

Code examples

You can find completed snippets of main.cpp and Platformio.ini at https://github.com/PacktPublishing/Accelerating-IoT-Development-with-ChatGPT/tree/main/Chapter_12.

In that updated main.cpp, you can see three new functions created by ChatGPT based on the previous version:

- connectToWiFi(): The function to connect Wi-Fi by SSID and password

- pingHost(): Ping google.com to check whether the public internet is accessible

- syncNTP: Obtain the current date and time from the public internet NTP server

In addition, in the updated platformio.ini, we also adopt a flexible approach to store and pass the following critical parameters to the main code using build_flags. When prompting, you can ask ChatGPT to use the same name macros from platformio.ini. Doing so would ensure that the output of ChatGPT does not use different names:

```
build_flags =
.......
-D WIFI_SSID=\"WiFi_TEST\"    //replace it by your actual Wi-Fi SSID
-D WIFI_PASSWORD=\"WiFi_Password\" //replace it by your actual Wi-
Fi password
-D PING_HOST=\www.google.com\    //the DNS server for ping destination
```

```
-D NTP_SERVER=\"pool.ntp.org\" //
the NTP server for time synchronization
-D GMT_OFFSET_SEC=-28800 //replace it by your local time zone, i.e.,
-28800 seconds means Pacific time zone, You can find your local time
zone information at https://en.wikipedia.org/wiki/List_of_UTC_offsets.
-D DST_OFFSET_SEC=3600 // Day saving time is on
.......
```

Utilizing `build_flags` in this context offers several benefits. It encapsulates your credentials' parameters, keeping them separate from your code and configuration files, which enhances security by reducing the risk of credential leakage. This practice also enhances portability, allowing you to easily transfer your project to another machine or development environment without exposing sensitive information. Updates become easier as well, as you only need to modify the `build_flags` section in your platformio.ini file to change these parameters, eliminating the need to alter your code. Importantly, `build_flags` is a build-time construct, making it platform-agnostic and usable on different platforms without modifications.

In the `platformio.ini` file example, there is a new library under `lib_deps`, `marian-craciunescu/ESP32Ping@^1.7`. Before compiling the code, you need to manually install it from the PlatformIO libraries, as shown in the following screenshot.

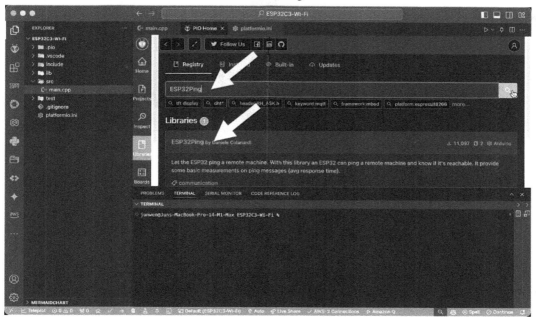

Figure 12.2 – Install the ESP32Ping library from PlatformIO

In this section, we asked ChatGPT to revise the previous code by adding a requirement for Wi-Fi access. Next, let's compile and upload the new version to ESP32 to verify the results.

Validating internet access on ESP32

In this section, we will observe the `Ping` command response and NTP synchronization result in the PlatformIO **TERMINAL** window. You can use PlatformIO to compile and upload this code to your ESP32 as per the steps in the last chapter.

Figure 12.3 shows a screenshot of the successful messages printed in the PlatformIO console **TERMINAL** window.

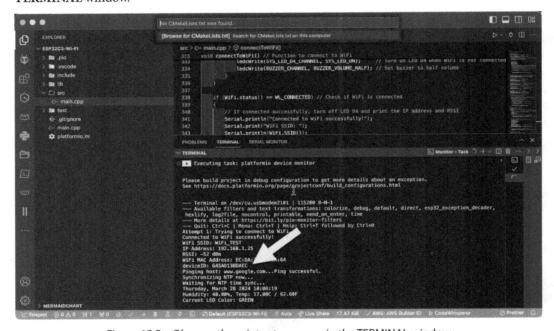

Figure 12.3 – Observe the printout message in the TERMINAL window

Make a note of the device ID displayed on your screenshot. The device ID is derived from the unique values read from the eFuse Mac address, using the `ESP.getEfuseMac()` function. We will use the device ID as the *THING*'s name when provisioning this device in AWS IoT Core in the next chapter. It's important to know that the eFuse Mac address cannot be altered once the hardware module has been manufactured.

In this section, you have learned using Platformio to validate the updated code generated by ChatGPT. As with *Chapter 11*, you might need to refine your prompts to ChatGPT to make them more specific. This iterative conversation will help fine-tune its code generation. After making adjustments, you should re-validate the new versions until the outcomes match your service flow diagram completely.

Summary

In this chapter, you connected the ESP32 to your local Wi-Fi network. You should be able to validate the preceding output in the PlatformIO console **TERMINAL** window, observing the Wi-Fi access, internet ping, and NTP synchronization information in addition to the temperature and humidity data.

In the next chapter, we will start the journey to connect the ESP32 to the AWS cloud, focusing on the IoT Core first. To do this, we will create a device in AWS IoT Core, generate access credential files, program these files in the code on ESP32, and set up a **TLS** (**Transport Layer Security**)-based **MQTT** (**Message Queuing Telemetry Transport**) connection between ESP32 and AWS.

13

Connecting the ESP32 to AWS IoT Core

In the previous chapter, we successfully instructed ChatGPT to generate code on the ESP32 to connect to your Wi-Fi network, obtain an IP address, ping an internet host, and synchronize with an **NTP (Network Time Protocol)** server to get the local time Now, we're moving on to the most critical step: connecting the ESP32 to **AWS IoT Core** via the MQTT protocol over a **Transport Layer Security (TLS)** connection.

By the end of this chapter, you will be able to possess the skills and knowledge to provision your ESP32 as a new **THING** in AWS IoT Core. Not only that, but you will also be able to connect your ESP32 to AWS IoT Core through a secured TLS/MQTT connection.

This chapter will cover the following topics:

- *Understanding the approach to connect the ESP32 to AWS IoT Core*: Learn about the X.509 certificates-based TLS-secured access mechanism implemented by AWS IoT Core

- *Creating a Thing in AWS IoT Core*: Provision your ESP32 as a new thing in the AWS IoT Core console, generate device certificates and private keys, and attach a connection policy

- *Creating AWS credential header file on the ESP32*: Generate a header file to store the contents of the AWS root **certificate authority (CA)** certificate, device certificates, and private key for TLS setup

- *Instructing ChatGPT to produce TLS and MQTT code on the ESP32*: Request ChatGPT to generate code by importing MQTT and TLS libraries and header files on the ESP32

- *Validating access status on the ESP32*: Monitor the connection status in the ESP32 local terminal window

Technical requirements

In this chapter, we will continue to use the `Platformio.ini` file to store and pass AWS IoT Core access information to the ESP32's code. In addition to creating a certificate header file, we will create a hardware information header file to identify the module information of the ESP32 and include it in the main code for device identification.

Understanding the approach to connect the ESP32 to AWS IoT Core

The IoT device access service provided by the AWS cloud is AWS IoT Core, as discussed in *Chapter 5* in the *AWS for IoT* section. As highlighted in that section, as of December 2023, the following four communication protocols are supported:

- **MQTT**
- **MQTT over WebSockets Secure (WSS)**
- **Hypertext Transfer Protocol – Secure (HTTPS)**
- **LoRaWAN**

Among these protocols, MQTT stands out as one of the most widely used. It accommodates the clients by enabling the publishing and subscribing of messages. The mechanisms of MQTT are explained in the *Chapter 5* section *Communication protocols between IoT devices and the cloud*.

Before establishing MQTT communication between an IoT device and AWS IoT Core, a secured low-layer connection must be created using TLS. TLS is a cryptographic protocol designed to provide secure communication over a network, supporting both server and client authentication.

Implementing TLS functionality on the ESP32 requires three credential files to establish an X.509 certificate-based TLS connection with AWS IoT Core: the AWS root CA file, your ESP32 device certificate `.pem` file, and the private key `.pem` file. The following are the explanations, their functions, and how to get them:

- **AWS root CA certificate**: When an IoT device tries to connect to AWS IoT Core, AWS IoT Core sends an X.509 certificate that the device uses to authenticate the server. Authentication occurs at the TLS layer through the validation of the X.509 certificate chain. This method is identical to the one your browser uses when visiting an HTTPS URL.

  AWS IoT Core supports two different data endpoint types, `iot:Data` and `iot:Data-ATS`. The `iot:Data` endpoints present a certificate signed by the VeriSign Class 3 Public Primary G5 root CA certificate. The `iot:Data-ATS` endpoints present a server certificate signed by an Amazon Trust Services CA. You can find the detailed information at `https://docs.aws.amazon.com/iot/latest/developerguide/server-authentication.html`.

Please note that the Arduino TLS library only supports the endpoint type of `iot:Data`. We will select the AWS IoT Core regions supporting the `iot:Data` type of endpoint, such as `us-west-2-amazonaws.com`. You can find those regions' capability at `https://docs.aws.amazon.com/general/latest/gr/iot-core.html`, in *Chapter 5* in the *AWS IoT Core* section.

You can download the AWS root CA certificate, VeriSign Class 3 Public Primary G5 root CA certificate for the `iot:Data` type endpoints, from `https://cacerts.digicert.com/pca3-g5.crt.pem`.

- **Device certificate**: In the context of TLS, a device certificate PEM refers to a digital certificate in **Privacy-Enhanced Mail (PEM)** format that a device uses to authenticate itself and establish a secure connection. The PEM format is a way to encode the certificate, which includes the following:

  - **A header**: `----BEGIN CERTIFICATE-----`
  - The base64-encoded **Distinguished Encoding Rules (DER)** certificate
  - **A footer**: `----END CERTIFICATE-----`

  This certificate contains the public key of the device and is signed by a CA. It's used in the TLS handshake process to prove the device's identity to another party.

- **Device private key**: In TLS, a device private key is also a critical component of the encryption process. It's a secret key that is generated by the device and never shared with anyone else. The device private key must be kept secure and confidential because anyone with access to it can decrypt the communication intended for the device. Like the device certificate PEM, the private key is also stored in a file, often in PEM format, and includes the following:

  - **A header**: `----BEGIN PRIVATE KEY-----`
  - The base64-encoded key
  - **A footer**: `----END PRIVATE KEY-----`

During the TLS handshake, the device does not transmit its private key. Instead, it uses the key to create digital signatures or to decrypt information sent to it that was encrypted with its public key, proving that it holds the corresponding private key without revealing it.

When you create a thing in the AWS IoT Core console, you will be prompted to download the device certificate PEM file and device private key file. You need to save them, copy and paste their contents into the credential file, which we will create in this chapter's exercise, and then program them to the ESP32. We will show you how to create the credential file and call it in the main code in this chapter.

In the ESP32 code implementation, we will use two standard Arduino libraries: `WiFiClientSecure.h` for establishing a TLS connection, and `PubSubClient.h` for performing MQTT communication. These two libraries are commonly used together to ensure secure MQTT communication over TLS:

- `WiFiClientSecure.h`: Part of the Arduino Wi-Fi shield library, this library supports TLS/SSL connections. It is a variant of `WiFiClient` designed to handle encrypted connections, vital for securely transmitting sensitive data over the Internet.

- `PubSubClient.h`: This library provides a client for simple publish/subscribe messaging with an MQTT server.

The following is the procedure to set up a TLS connection between the ESP32 and AWS IoT Core:

1. Log in to the AWS console.

2. Select **AWS IoT Core service**

3. Create a *Thing*.

4. Generate and download certificates (device certificate PEM and private key).

5. Generate policy and attach it to certificates.

6. Create `SecureCredentials.h` in PlatformIO by the AWS root CA certificate, and the downloaded certificates.

7. Program code on the ESP32 by calling `WiFiClientSecure.h` and `PubSubClient.h` to set up a TLS/MQTT connection.

8. Observe print messages on the ESP32 local terminal window.

In this section, we went through the entire procedure, from provisioning a device in AWS IoT Core to establishing a TLS connection. In the next section, we will complete the procedure step by step.

Provisioning the ESP32 in AWS IoT Core

Before starting provisioning the ESP32 in AWS IoT Core, you must complete the following tasks as prerequisites:

- Create an AWS account in your region

- Have an admin role with a permission policy of **AdministratorAccess**

Assuming you already completed the preceding tasks, let us follow these steps to provision your ESP32 in AWS IoT Core:

1. Login to your AWS account from AWS Console as an **IAM (Identity and Access Management)** user by your admin role:

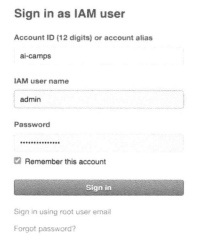

Figure 13.1 – AWS login page

2. Find and click the **IAM** service:

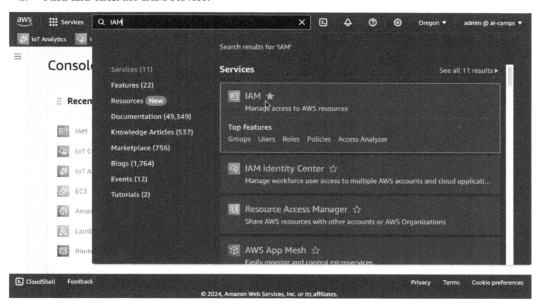

Figure 13.2 – Locate and start the IAM console

3. Find and click on **Users** on the left side:

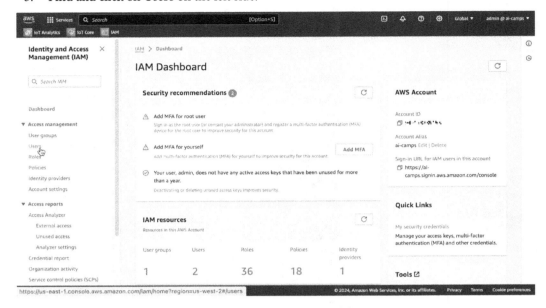

Figure 13.3 – Locate Users in the IAM console

4. Find the regular IAM user for accessing AWS IoT Core service, such as jun.wen in the following screenshot. If you don't have such a regular IAM user, you can click **Create User** to set up one in the following screenshot. This operation adheres to AWS's *least privilege* principle, a fundamental security practice that involves granting users and systems the minimal access levels needed for their tasks. Implementing this principle in AWS is essential for minimizing potential security risks and ensuring the security of your AWS environment. In this step, we will create a regular user and grant them sufficient access to eligible AWS resources. Once we complete this operation, we will use this regular IAM user to execute the configuration from the rest of the steps.

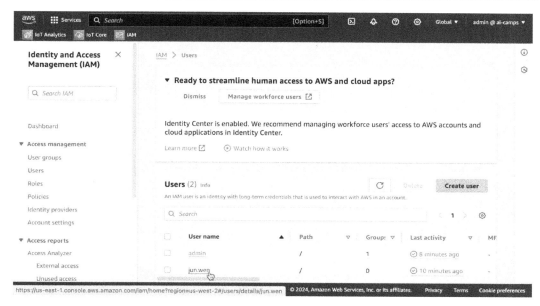

Figure 13.4 – Locate a regular IAM user

5.   Click the regular IAM user name, and make sure that it is attached to these five permissions policies as in the following screenshot since we are going to use all these services in later steps:

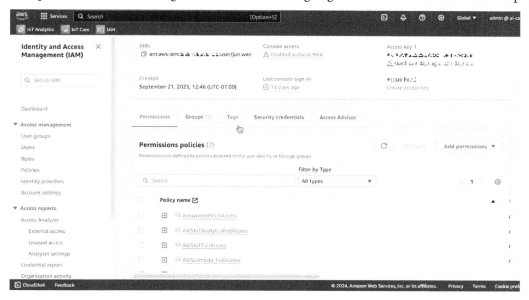

Figure 13.5 – Check the permission policies of this regular IAM user

In this step, we will attach this regular IAM user to the following permission policies needed by this chapter and *Chapter 15*:

- `AWSIoTFullAccess`: This gives access to IoT full services and will be used in this chapter.

- `AmazonSNSFullAccess`: This gives access to the **SNS (Simple Notification Service)** service and will be used in *Chapter 15*.

- `AWSIoTAnalyticsFullAccess`: This gives access to the IoT Analytics service and will be used in *Chapter 15*.

- `AWSLambda_FullAccess`: This gives access to the Lambda service and will be used in *Chapter 15*.

6. If you didn't find these permissions policies attached to this regular IAM user, please click **Add permission** in the preceding screenshot. Search, and attach policies directly as the following screenshot:

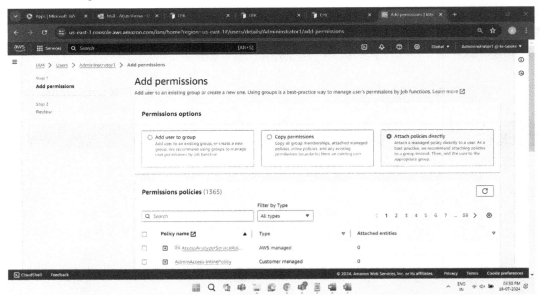

Figure 13.6 – Attach the necessary permissions to this regular IAM user

7. Now, you are all set to log in to the AWS console as this regular IAM user as follows:

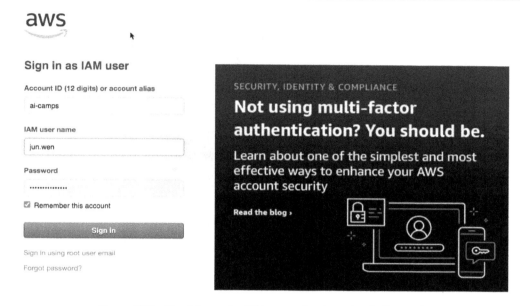

Figure 13.7 – Use this regular IAM user to log in to the AWS console

After login, select the region close to your location and support the endpoint type of `iot:Data`, such as `us-west-2 (iot.us-west-2.amazonaws.com)`; you can find it at `https://docs.aws.amazon.com/general/latest/gr/iot-core.html` in **AWS IoT Core - control plane endpoints**:

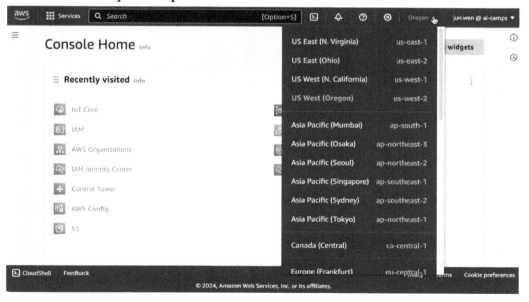

Figure 13.8 – Select your region with the AWS IoT Core service

8. Find and click the **IoT Core** service:

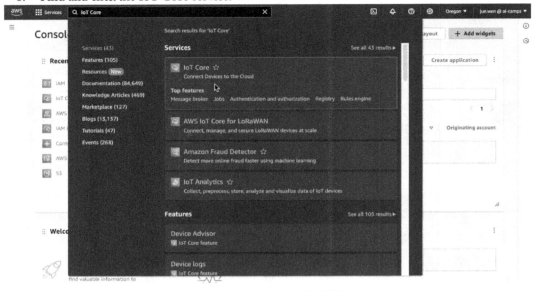

Figure 13.9 – Locate and move to the IoT Core console

9. You will see the **AWS IoT** service as it is in the following screenshot:

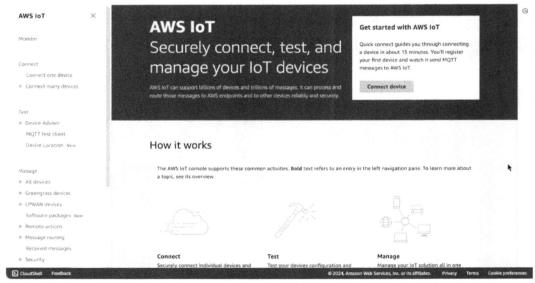

Figure 13.10 – The IoT Core welcome page

10. Find and click **Things** on the left side and then click **Create things**:

Figure 13.11 – Start to create things

11. Click **Create single thing** and click on **Next**:

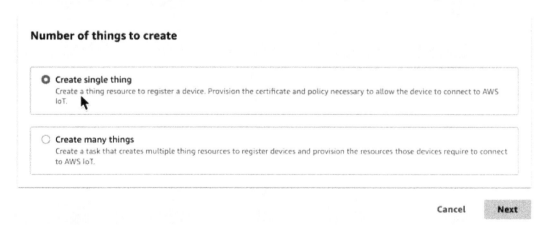

Figure 13.12 – Select Create single thing

12. Give the thing a `deviceID` name, the eFuse MAC you got from running the code in the last chapter.

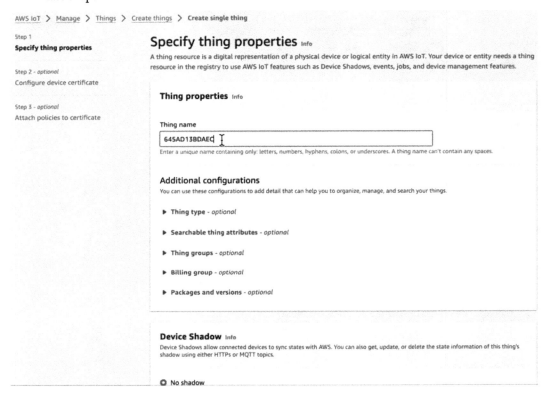

Figure 13.13 – Give deviceID to Thing name

13. Generate the device certificates. Please note the certificate files can be shared with other devices. When you create additional devices after the first one, you can select **Skip creating a certificate at this time**, and then attach the same certificates to these devices:

Figure 13.14 – Generate device certificate

14. Now, you can download all these files for future usage, or just download **Device certificate** and **Private key file** as shown in the following screenshot. Save these two files on your computer, and we will copy and paste their contents to the credential header file to program on the ESP32.

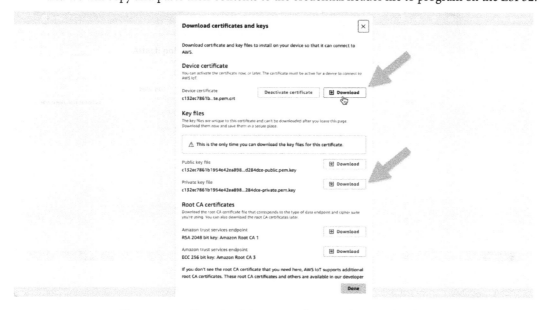

Figure 13.15 – Download device certificate and private key file

15. Click **Create thing** to complete the first device creation process.

Figure 13.16 – Complete the new device creation

16. You will see the **successfully created** message as shown in the following screenshot and the first device with the thing's name of `deviceID` that you previously gave.

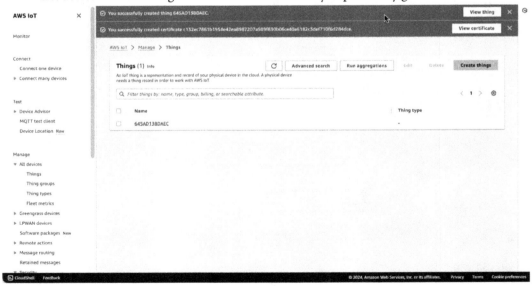

Figure 13.17 – The successful page

17. On the left side, find and click **Policies** to create a policy:

Figure 13.18 – Create IoT Core access policy

18. You can enter a policy name here, for example, `AWS_IOT_Core_Access`, as shown in the following screenshot. Then, click **JSON**, and copy and paste the following code to the policy document. Please note that you must use your actual region (for example, `us-west-2`) and your account ID. You can get your account ID information by clicking your account name in the top right corner:

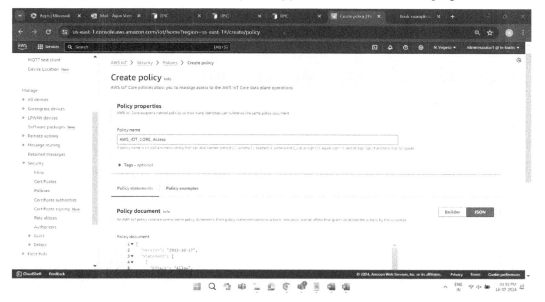

Figure 13.19 – Edit the policy document

Here is a closer look at the contents of the policy document (you can find the example at `https://github.com/PacktPublishing/Accelerating-IoT-Development-with-ChatGPT/blob/main/Chapter_13/AWS_IOT_Core_Access`):

```
1.  {
2.    "Version": "2012-10-17",
3.    "Statement": [
4.      {
5.        "Effect": "Allow",
6.        "Action": "iot:Connect",
7.        "Resource": "arn:aws:iot:your_region:your_account_
id:client/*"
8.      },
9.      {
10.       "Effect": "Allow",
11.       "Action": "iot:Publish",
12.       "Resource": "arn:aws:iot:your_region:your_account_
id:topic/*"
13.     },
14.     {
15.       "Effect": "Allow",
16.       "Action": "iot:Subscribe",
17.       "Resource": "arn:aws:iot:your_region:your_account_
id:topicfilter/*"
18.     },
19.     {
20.       "Effect": "Allow",
21.       "Action": "iot:Receive",
22.       "Resource": "arn:aws:iot:your_region:your_account_
id:topic/*"
23.     }
24.   ]
25. }
26.
```

19. Now, you will create a policy called `AWS_IOT_Core_Access` as shown in the following screenshot:

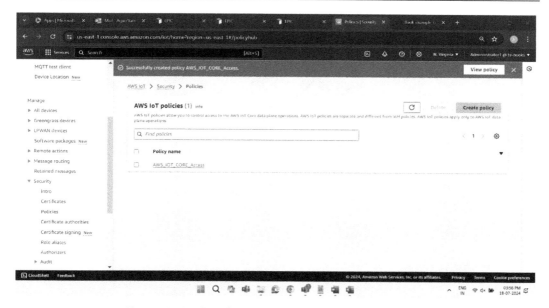

Figure 13.20 – Complete AWS IoT Core access policy creation

20. On the left side, click **Certificate** and you will see the certificate ID that has been created. Click this **Certificate ID**:

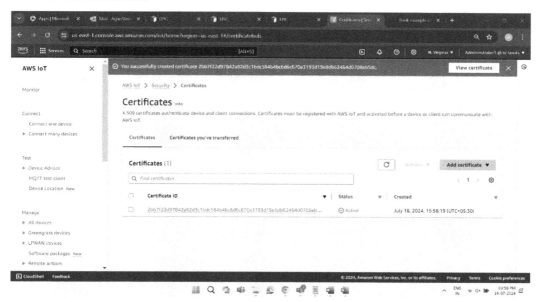

Figure 13.21 – Locate certificates previously created

21. Now, click **Attach policies**:

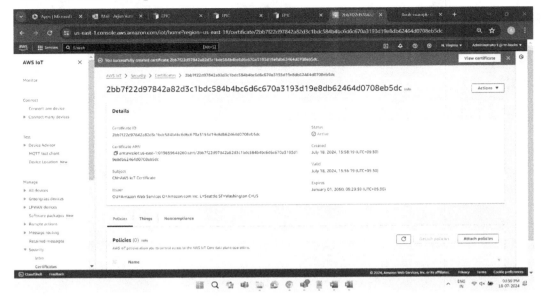

Figure 13.22 – Start to attach a policy to the certificate

22. In the prompt window, select the policy we created in the previous step.

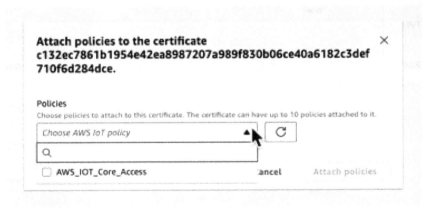

Figure 13.23 – Locate the policy previously created

23. Now, you can see the policy that is attached to this certificate ID.

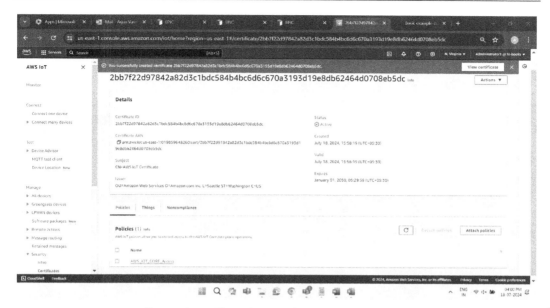

Figure 13.24 – The policy attachment successful page

Now, you have completed a device provisioning process in AWS IoT Core including the following:

- You have created a device with a thing name of the device ID of your ESP32 (the device's eFuse MAC address)
- You have generated the device certificates and private key
- You have downloaded the device certificates and private key on your PC
- You have created a connection policy and attached it to the certificate ID

You can use the same certificates, private keys, and connection policies that you create for other devices in the future. This means that you don't need to create new certificates, keys, and policies when you add more devices. In this step, you have completed your ESP32 provisioning in AWS IoT Core, generated a device certificate and a private key, and created a permission policy to access AWS IoT Core. The configuration task was done on the AWS side in this chapter; now, let us move on to programming AWS to support TLS and MQTT.

Creating an AWS credential header file on the ESP32

As stated, we will make a credential header file that contains the AWS root CA certificate, device certificate, and private key, which is requested for TLS setup. This header file will be imported and called by the main code on the ESP32.

You can find an example of the SecureCredential.h header file at https://github.com/ PacktPublishing/Accelerating-IoT-Development-with-ChatGPT/tree/main/ Chapter_13. This header file will be called in main.cpp. Your SecureCredential.h contents shall look like the following screenshot.

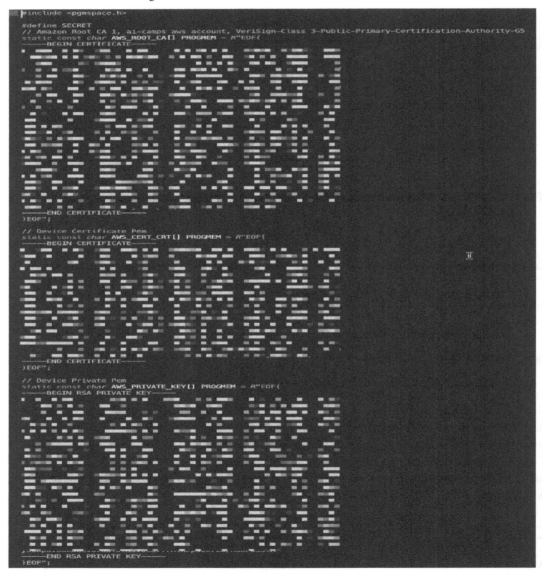

Figure 13.25 – The SecureCredential.h file format example

In addition to the credential header file, we will have to create a hardware header file to read the ESP32 chipset hardware information, including the eFuse MAC address for `deviceID`. You can find an example of the `HardwareInfo.h` header file in the same link folder.

As long as you created these two header files, `SecureCredential.h` and `HardwareInfo.h`, in the next step, we are going to continue to prompt ChatGPT to update the code from the previous chapter to call these new header files, `SecureCredential.h`, `HardwareInfo.h`, `WiFiClientSecure.h`, and `PubSubClient.h`, in the `mail.cpp` code, and initiate TLS/MQTT connection with AWS IoT Core.

Instructing ChatGPT to produce TLS code on the ESP32

Now, let us list the necessary information in the ESP32's main code to access AWS IoT Core:

- TLS certificates, which will be stored in the `SecureCredentials.h` header file
- MQTT server address and port, which will be stored in the `Platformio.ini` file, for example, `AWS_IOT_MQTT_SERVER=\"xxxxxxxxxxxxxxx.iot.your_aws_region.amazonaws.com\"`, `AWS_IOT_MQTT_PORT=8883`
- `deviceID`, which will be populated from the `HardwareInfo.h` header file

Be aware that `AWS_IOT_MQTT_SERVER` is the endpoint address as shown in the following figure; it is in the settings under AWS IoT. Please remember to remove `-ats` from the endpoint address because we use the endpoint type of `iot:Data`, not `iot:Data-ATS`.

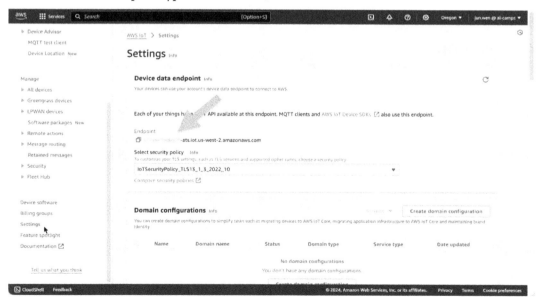

Figure 13.26 – Locate your endpoint information

You can now continue using the code created in the previous chapter and add the TLS MQTT requirement as the following prompt:

Hi, ChatGPT.

Please update the previous code, maintain its current structure, style, and output format, and support the following additional requirements:

1. Import and use the ESP32 standard libraries of WiFiClientSecure.h for TLS and PubSubClient.h for the MQTT stack.

2. Import SecureCredentials.h, which stores AWS certificates, and call it in the main code.

3. Import HardwareInfo.h to read the eFuse MAC address as the deviceID.

4. Store AWS_IOT_MQTT_SERVER and AWS_IOT_MQTT_PORT information in the build_flagsin Platformio.ini file.

5. Create a dedicated function called connectAWS in the main code for AWS IoT Core access.

6. Normal condition: If the ESP32 successfully accesses AWS IoT Core, print out successful messages.

7. Abnormal condition: If the ESP32 fails to access AWS IoT Core, print out the received error messages.

Code examples

There is a sample of the main.cpp code, which is located at https://github.com/PacktPublishing/Accelerating-IoT-Development-with-ChatGPT/tree/main/Chapter_13.

In the main.cpp code, you'll find a new function called connectAWS(), created by ChatGPT. This function is located in the setup() section and is responsible for initiating the TLS connection. It calls the root CA, device certificate, and private key contents during this process:

```
void connectAWS() // Function to connect to AWS IoT Core
{
    // Configure WiFiClientSecure to use the AWS IoT device
credentials
    net.setCACert(AWS_ROOT_
CA);          // Set the AWS Root CA certificate
    net.setCertificate(AWS_CERT_CRT);   // Set the device certificate
    net.setPrivateKey(AWS_PRIVATE_KEY); // Set the private key
```

```
    // Set the AWS IoT endpoint and port
    mqttClient.setServer(AWS_IOT_MQTT_SERVER, AWS_IOT_MQTT_PORT);

    Serial.println("Connecting to AWS IOT Core");

    while (!mqttClient.connect(deviceID.c_
str())) // Connect to AWS IoT Core
    {
        Serial.print(".");
        delay(MQTT_RECONNECT_DELAY_MS);
    }

    if (!mqttClient.connected()) // Check if the client is connected
    {
        Serial.println("AWS IoT Core connection is failed!");
        return;
    }
    Serial.println("AWS IoT Core is connected successfully!");
}
```

Before you compile and upload the new code, you need to search and install `PubSubClient.h` from PlatformIO **Libraries**; please refer to the following screen to execute it.

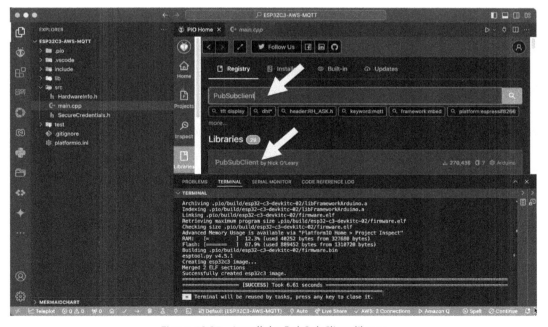

Figure 13.27 – Install the PubSubClient library

You now save the `SecureCredentials.h` and `HardwareInfo.h` files in the same folder in PlatformIO with the `main.cpp` code, as shown in the following screenshot:

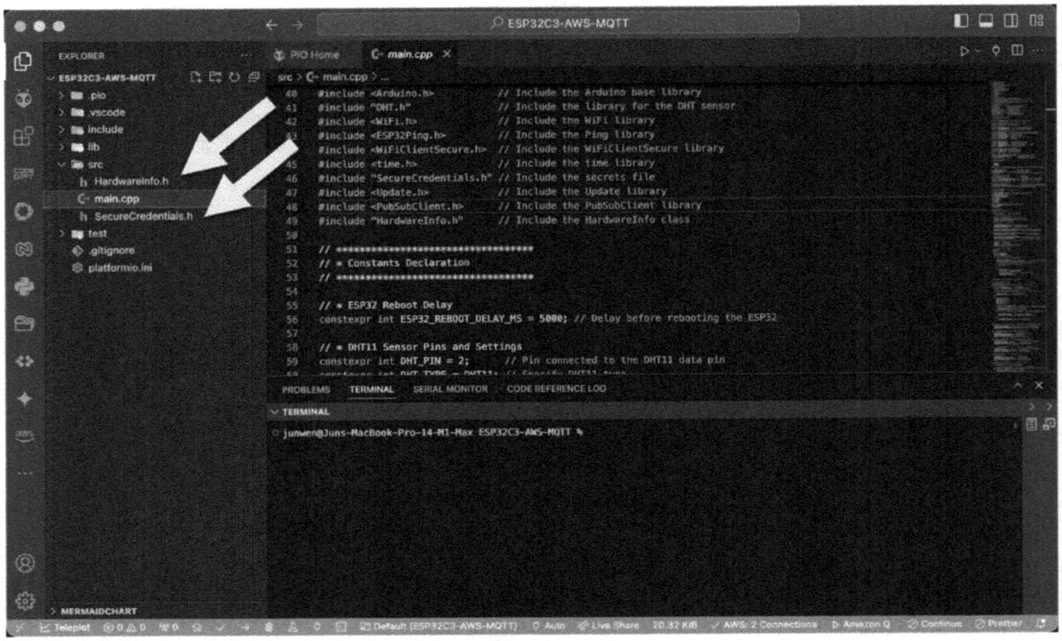

Figure 13.28 – Copy SecureCredential.h and HardwareInfo.h in the folder with main.cpp

By the end of this section, you will be able to request ChatGPT to update the code to support a TLS/MQTT connection. In the next section, we will compile and upload the code and validate the access status on the terminal window of the ESP32.

Validating access status on the ESP32

You can now copy and paste the updated main code from ChatGPT, compile and transfer it to the ESP32, and check the access status in the local terminal window, as shown in the following screenshot.

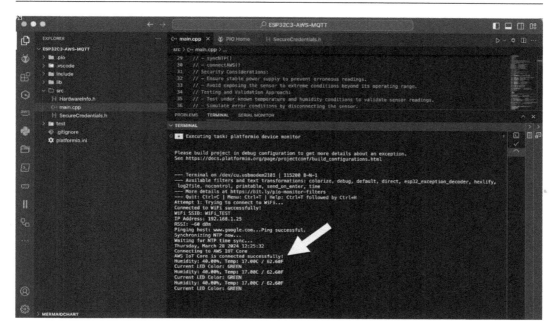

Figure 13.29 – Validate AWS IoT Core connection status

In this section, you uploaded the revised version of the `main.cpp` code to the ESP32. After that, you will check the printout message regarding the connection status between the ESP32 and AWS IoT Core. If your `SecureCredential.h` header file and AWS MQTT server are configured correctly, your ESP32 should successfully connect to AWS IoT Core.

Summary

In this chapter, you made the ESP32 access to the AWS cloud through MQTT over a TLS connection. This is the fundamental step to send sensor data to the AWS cloud.

In the next chapter, after a successful connection to AWS IoT Core, we will use the Arduino JSON library on the ESP32 to publish DHT11 temperature and humidity data to AWS, and we will observe the received MQTT message at AWS IoT Core. We will continue to use ChatGPT to build the code on the ESP32 to publish the MQTT topic and create a JSON payload.

14

Publishing Sensor Data to AWS IoT Core

In the previous chapter, we successfully instructed ChatGPT to generate code to connect the ESP32 to AWS IoT Core through a TLS/MQTT secured connection. In this chapter, we are going to use the MQTT protocol to publish sensor data to the AWS cloud.

By the end of this chapter, you will be able to instruct ChatGPT to program code on ESP32, create an MQTT Publish topic, and generate a **JavaScript Object Notation** (**JSON**) document. This will allow you to encapsulate sensor data for delivery to AWS IoT Core.

This chapter will cover the following topics:

- *Understanding the approach to sending sensor data through MQTT Publish*: Learn about the data delivery mechanism of the MQTT Publish operation

- *Constructing MQTT publish topic and payload in ESP32*: Instruct ChatGPT to generate code to create an MQTT topic and JSON payload

- *Validating the delivered sensor data*: Monitor the execution result in a local terminal window and observe the received messages from the AWS MQTT test client

Technical requirements

In this chapter, we will instruct ChatGPT to update the code on the ESP32 that we created in *Chapter 13*. On the AWS cloud, you will log in to AWS IoT Core to configure an MQTT test client to subscribe to a topic, which is the MQTT publish topic you created in the ESP32 code.

In this book, we will only illustrate the upstreaming operation from the DHT11 sensor to its report data to AWS IoT Core using MQTT Publish. The operation of transferring data from AWS IoT Core to an IoT device is called MQTT Subscribe. This operation, used for example when performing a device firmware upgrade, will not be used in our project.

Sending sensor data through MQTT Publish

In *Chapter 5*, we discussed the MQTT protocol and its Publish/Subscribe model. The following figure describes how an MQTT broker (such as AWS IoT Core) interacts with an MQTT Publisher or an MQTT Subscriber (such as IoT devices). In the previous chapter, we established a TLS connection between the MQTT Publisher (ESP32) and the MQTT broker (AWS IoT Core).

Once the TLS connection is successfully set up, the next step is for the IoT device, the MQTT Publisher, to encapsulate sensor payload data in JSON format with a declared *topic* to send to the MQTT broker, AWS IoT Core.

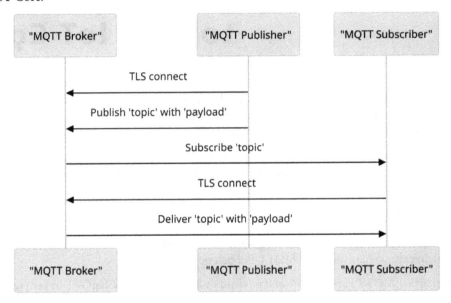

Figure 14.1 – MQTT flow

The following definitions pertain to *topic*, *payload*, and *JSON*, terms used in the MQTT protocol flow:

- **Topic**: This is typically a string. In the case of our project, it is formatted as `deviceID/pub` to distinguish the topic amongst various devices. The unique `deviceID` value is derived from `ESP.getEfuseMac()`, which is included in the `HardwareInfo.h` header file created in the previous chapter.

- **Payload**: This term refers to sensor raw data; in the case of this project, it includes things such as device information and temperature and humidity values, which are formatted using JSON.

  JSON is a data interchange format that is lightweight in nature and highly user-friendly. It is designed to be easily read and written by humans while simultaneously being simple enough for machines to parse and generate. JSON is not tied down to any specific programming language.

Instead, it is a language-agnostic text format that has been widely accepted and adopted across a variety of languages. These languages include, but are not limited to, C, C++, C#, Java, JavaScript, Perl, and Python. In terms of structure, JSON is composed of different forms. An object, for instance, is an unordered collection of name/value pairs. The beginning of an object is signaled by a left brace (*{*), and it concludes with a right brace (*}*). Within the object, each name is succeeded by a colon, and the name/value pairs are differentiated by a comma.

Let's consider the following example to clarify this concept:

```
1.  {
2.    "deviceInfo": {
3.      "macAddress": "00:1A:2B:3C:4D:5E",
4.      "firmwareVersion": "1.0.5"
5.    },
6.    "sensorData": {
7.      "temperatureC": 22.5,
8.      "temperatureF": 72.5,
9.      "humidity": 50
10.   }
11. }
12.
```

- **Serialization**: This refers to the process where JSON format data is written to a serialized string. This string is then encapsulated within an MQTT Publish message. This specific procedure is known as the `JsonSerializer.Serialize` method. This method is an essential aspect of data transmission and storage, as it allows complex data structures to be converted into a format that can be easily stored or transmitted, and then reconstructed when necessary. By utilizing the `JsonSerializer.Serialize` method, it is possible to effectively manage and manipulate JSON data, facilitating seamless data exchange and communication. In the PlatformIO IDE, there is a standard library, `ArduinoJson.h`, that is used to perform JSON format data serialization.

In this section, we explored the key terms in MQTT communication. These terms are essential for exchanging information between IoT devices and the MQTT server. In the next section, we will apply these terms to our project.

Constructing an MQTT Publish topic and payload in ESP32

In the previous chapter, we introduced two standard Arduino libraries related to TLS/MQTT functionality. The `WiFiClientSecure.h` library is used to set up a TLS connection, while the `PubSubClient.h` library is used for MQTT publish.

In this chapter, we will introduce another Arduino library, `ArduinoJson.h`. This library is used to construct a JSON document and serialize it for transmission to AWS IoT Core.

ArduinoJson.h is a versatile C++ library used for parsing, serializing, and manipulating JSON data on microcontrollers such as the ESP32, making it ideal for IoT applications. It's designed with memory efficiency, ease of use, and flexibility in mind, making it a popular choice among developers. To use it in a PlatformIO project, include it in your `platformio.ini` file under `lib_deps`. Then, incorporate it into your source code with `#include <ArduinoJson.h>`.

We will manually install this library in Platform before we compile the `main.cpp` code.

Based on the code created in the previous chapter, you can continue to prompt ChatGPT to add an MQTT Publish topic, the JSON data format, and the serialization requirement with the following prompt:

```
Hi, ChatGPT
```

```
Please update the previous code, maintain its current structure, style
and output format, and support the following additional requirements:
```

- Import and use ArduinoJson.h for JSON data format and serialization
  operation.

- Import SecureCredentials.h which stores AWS certificates and
  call it in main code.

- Import HardwareInfo.h to read eFuse MAC as the unique deviceID.

- Build the publish topic AWS_IOT_PUBLISH_TOPIC by deviceID + "/
  pub".

- Create a dedicated function "mqttPublishMessage" to send sensor
  data.

- Create a JSON string including objects of "timeStamp", "deviceID",
  "status", "date", "time", "timezone", "DTS", "data", and values
  for "data" with "temp_C", "temp_F" and "humidity".

Next, let us look at some code examples.

Code examples

You can find a sample of the `main.cpp` code at `https://github.com/PacktPublishing/Accelerating-IoT-Development-with-ChatGPT/tree/main/Chapter_14`. The following is an extract `main.cpp` code. ChatGPT created a new function as requested, `mqttPublishMessages()`:

```cpp
void mqttPublishMessage(float humidity, float temperatureC, float tem-
peratureF, SensorConditionStatus condition) // Function to pub-
lish message to AWS IoT Core
{
    if (!mqttClient.connected()) // Check if the client is connected
```

```
    {
        connectAWS(); // Connect to AWS IoT Core if not connected
    }
    // Fetch the current time
    struct tm timeinfo;
    if (!getLocalTime(&timeinfo))
    {
        Serial.println("Failed to obtain time");
        return; // Don't proceed if time couldn't be obtained
    }

    // Format the date and time separately
    char formattedDate[11]; // Buffer to hold the formatted date "mm-
dd-yyyy"
    char formattedTime[9];   // Buffer to hold the format-
ted time "hh:mm:ss"

    // Use strftime to format the date and time separately
    strftime(formattedDate, sizeof(formattedDate), "%m-%d-
%Y", &timeinfo);
    strftime(formattedTime, sizeof(formatted-
Time), "%H:%M:%S", &timeinfo);

    // Get Unix time
    time_t unixTime = mktime(&timeinfo);

    String conditionStr = condition == Normal ? "Normal" : condi-
tion == BelowNormal ? "Below Normal"

:condition == AboveNormal    ? "Above Normal"

: "Sensor Error";

    // Create a JSON document
    StaticJsonDocument<256> doc;

    // Populate document
    doc["timeStamp"] = unixTime;
    doc["deviceModel"] = "DHT11";
    doc["deviceID"] = String(ESP.getEfuseMac(), HEX);
    doc["status"] = conditionStr;
    doc["date"] = formattedDate;
    doc["time"] = formattedTime;
    doc["timeZone"] = timezoneStr;
    doc["DST"] = dstStatus;
```

```
    JsonObject data = doc.createNestedObject("data"); // Cre-
ate a nested object for data
    data["temp_C"] = temperatureC;
    data["temp_F"] = round(temperatureF);
    data["humidity"] = humidity;

    String jsonString;                      // Cre-
ate a string to hold the JSON data
    serializeJson(doc, jsonString);         // Serialize the JSON docu-
ment to a string
    Serial.print("Publishing message: "); // Print the message
    Serial.println(jsonString);            // Print the JSON data

    // Determine buffer size
    size_t jsonSize = measureJson(doc) + 1; // +1 for null terminator
    Serial.print("Calculated JSON buffer size: ");
    Serial.println(jsonSize); // Print the buffer size

    if (!mqttClient.publish(AWS_IOT_PUBLISH_TOPIC.c_str(), json-
String.c_str()))
    {
        Serial.println("Publish failed");
    }
    else
    {
        Serial.println("Publish succeeded");
    }
}
```

This function publishes sensor data to AWS IoT Core using the MQTT protocol. It takes in four parameters: humidity, temperatureC (in Celsius), temperatureF (in Fahrenheit), and condition. These represent the sensor's humidity, temperature readings, and condition, respectively.

The function first checks if the MQTT client is connected to AWS IoT Core. If not, it calls the connectAWS function to establish a connection.

It then obtains the current time using the getLocalTime function. If unsuccessful, it prints an error message and exits the function. If successful, it formats the date and time separately with the strftime function.

Next, it creates a conditionStr string to display the sensor condition in a human-readable format. It does this by using a ternary operator to map the SensorConditionStatus enum to a string.

The function then constructs a JSON document using the StaticJsonDocument class from the ArduinoJson library. This document includes various details such as the timestamp, device model, device ID, sensor status, date, time, timezone, daylight-saving time status, and sensor data (temperature and humidity).

The JSON document is serialized to a string using the serializeJson function, and this string is printed to the serial monitor for debugging.

The function calculates the size of the JSON document with the measureJson function and prints this size to the serial monitor.

Finally, the function tries to publish the JSON string to AWS IoT Core using the mqttClient. publish function. If unsuccessful, it prints an error message. If successful, it prints a success message.

In the same folder of https://github.com/PacktPublishing/Accelerating-IoT-Development-with-ChatGPT/tree/main/Chapter_14, you also can see an example of the platformio.ini file, reproduced here as follows:

```
1.  [env:esp32-c3-devkitc-02]
2.  platform = espressif32
3.  board = esp32-c3-devkitc-02
4.  framework = arduino
5.  monitor_filters = esp32_exception_decoder, colorize
6.  monitor_speed = 115200
7.  build_src_filter = +<../../src/>  +<./>
8.  board_build.flash_mode = dio
9.  build_flags =
10.   -D ARDUINO_USB_MODE=1
11.   -D ARDUINO_USB_CDC_ON_BOOT=1
12.   -D WIFI_SSID=\"WiFi_SSID\"
13.   -D WIFI_PASSWORD=\"WiFi_Password\"
14.   -D PING_HOST=\"www.google.com\"
15.   -D NTP_SERVER=\"pool.ntp.org\"
16.   -D GMT_OFFSET_SEC=-28800
17.   -D DST_OFFSET_SEC=3600
18.   -D AWS_IOT_MQTT_SERVER=\"endpoint.iot.your_aws_iot_region.
amazonaws.com\"; please be noted the "endpoint" information can be
found at the "Settings" in AWS IoT.
19.   -D AWS_IOT_MQTT_PORT=8883
20.   -w
21.  lib_deps =
22.   adafruit/DHT sensor library@^1.4.6
23.   adafruit/Adafruit Unified Sensor@^1.1.14
24.   marian-craciunescu/ESP32Ping@^1.7
25.   bblanchon/ArduinoJson@^7.0.4
26.   knolleary/PubSubClient@^2.8
```

There is a new entry of **bblanchon/ArduinoJson** under **lib_deps**. You need to find and install it manually in Platformio libraries.

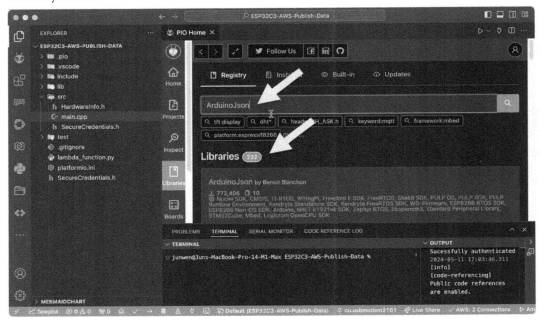

Figure 14.2 – Install the ArduinoJson library in PlatformIO

In this section, you have learned the key components of the MQTT protocol and instructed ChatGPT to update your previous code to support MQTT Publish operations by creating a new dedicated function. Now, let us compile the code, upload it, and validate the result.

Validating the delivered sensor data

After compiling and uploading the updated code to ESP32, you can check the execution result in the local terminal window.

In the following screenshot, the publishing message shown includes the serialized JSON format payload as defined in the main code.

Figure 14.3 – Validate the result in the local terminal window

Now you can login to the AWS Management Console and select **AWS IoT** as shown in the following screenshot. Under **Test**, click **MQTT test client**.

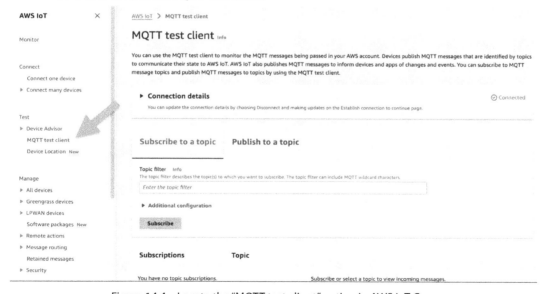

Figure 14.4 – Locate the "MQTT test client" option in AWS IoT Core

Under **Subscribe to a topic**, find the **Topic filter** field and type in +/pub. The + wildcard character matches exactly one item in the topic hierarchy. For example, a subscription to Sensor/+/room1 will receive messages published to Sensor/temp/room1, Sensor/moisture/room1, and so on.

In the code of ESP32, we defined the topic by **deviceID/pub**, thus +/pub here can match every device's MQTT publish topic. In the following screenshot, you can see the connection details read **Connected**, and the sensor data is in JSON format as we defined in the main code.

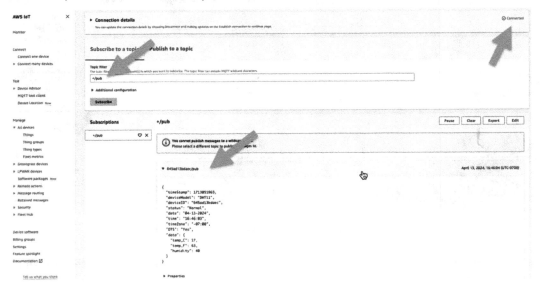

Figure 14.5 – Validate the published messages in AWS IoT Core

Now that we are at the end of this section, you will be able to observe the MQTT Publish message operation in both the local terminal window and the AWS IoT Core MQTT test client. If the results are not as expected, refine your prompts to more specifically suit your requirements.

Summary

At the end of this chapter, the sensor data has successfully reached AWS IoT Core. However, this is merely the initial step for a cloud service in commercial-grade IoT applications. The raw data now requires further processing. This can involve triggering an immediate alarm if an abnormal event is detected or storing the data for future analysis and visualization.

In the next chapter, you will learn storing, processing, and analyzing this sensor data using various AWS services. We plan to establish two message-routing rules in AWS IoT Core. The first rule will involve AWS Lambda, the premier serverless computing service. If abnormal sensor data is detected, we will request ChatGPT to generate a Python-based algorithm on AWS Lambda, which will trigger an email alert on AWS SNS. The second rule will guide data to AWS IoT Analytics for storage and preparation for subsequent visualization.

15

Processing, Storing, and Querying Sensor Data on AWS Cloud

In the previous chapter, we successfully instructed ChatGPT in creating an MQTT publish topic and a JSON-format payload in ESP32, which then delivered the serialized data to AWS IoT Core. Starting from this chapter, we will use additional AWS services. These will allow us to process abnormal events, trigger corresponding email alerts, and store, transform, sort, and query all the received sensor data.

By the end of this chapter, you will have acquired the necessary skills to configure a variety of other AWS services to process, store, and query the collected sensor data. This will allow you to carry out a range of critical tasks that are routinely employed in the context of commercial deployment.

In this chapter, we're going to cover the following main topics:

- Creating a customer-managed policy
- Task 1 – abnormal event process
- Task 2 – data storage and querying

Creating a customer-managed policy

In *Chapter 13*, using the principle of *least privilege* in AWS, we created a regular IAM user and attached the following permission policies to it:

- `AWSIoTFullAccess`
- `AmazonSNSFullAccess`
- `AWSIoTAnalyticsFullAccess`
- `AWSLambda_FullAccess`

These permissions are tagged as **AWS Managed** policies in the AWS IAM console, which are created and managed by AWS. They are designed to simplify the process of granting permissions for commonly used AWS services and tasks. AWS-managed policies ensure that your IAM roles adhere to best practices for security, aligning with AWS guidelines. However, for more specific use cases or to gain more granular control, usually you need to create some *customer-managed* policies.

Unlike *AWS-managed policies* predefined by AWS for use across multiple accounts, a customer-managed policy is an IAM policy you manage within your own AWS account. This gives you full control over the permissions you set, allowing for more detailed management of your AWS resources.

As we are going to use the necessary AWS resources—IoT Core, Lambda, SNS, and IoT Analytics—to complete the following two tasks, these services must be granted the proper role and permission to execute operations at our request.

Here are the configuration steps for policy settings for these two tasks:

1. Log in to the AWS IAM console using your admin role.
2. Under **Access management | Policies**, click **Create policy**.
3. In **Policy editor**, create the policy statement by clicking **JSON**.

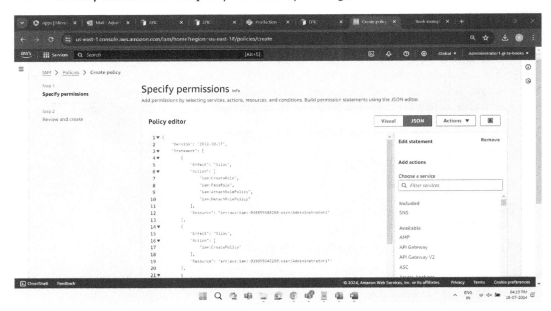

Figure 15.1 – Create policy statement

You can find an example policy statement file at https://github.com/PacktPublishing/Accelerating-IoT-Development-with-ChatGPT/blob/main/Chapter_15/aws_customer_managed_policy.

This customer-managed policy example grants sufficient permissions to set up the two tasks we are going to complete in this chapter. Here's a breakdown and explanation of each section:

- **IAM Role Management**: It permits reading, listing, creating, and passing IAM roles, limited to service roles within your own AWS account

- **IAM Policy Management**: It allows the listing, creation, attachment, and detachment of IAM policies for the resource under the same AWS account by the service role instead of identity-based policy

- **IAM User Management**: It provides permissions to list and get users, roles, policies, and their attachments

- **KMS Key Description**: It is required when the IAM user creates an SNS topic

- **SNS Publishing**: It grants permission to publish messages to any SNS topic in your region

- **Access Analyzer**: The final statement allows listing policy generations and validating policies using AWS IAM Access Analyzer across all resources

Please note that you need to replace `<your_region>` and `<your_account_id>` in this policy statement example with your own information. You can find the value for `<your_account_id>` showing under your admin account in the upper-right corner.

4. You can now name this policy and create it. For instance, you might call it `Service_Role_Policy_For_IoT`. However, feel free to choose a name that suits your preference.

5. In the **IAM** console, click on **Users**, select the desired IAM user, for instance, **jun.wen**, then click **Add Permissions | Attach policies directly**, search for `Service_Role_Policy_For_IoT`, and attach it to this IAM user. You should be able to see it under **Policy name**.

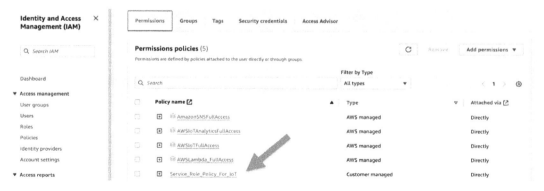

Figure 15.2 – New policy attached

Now, as an IAM admin, we have granted the IAM user the privilege required to execute the setup procedure for the following two tasks.

In the realm of commercial IoT applications, the arrival of data in the cloud always triggers two fundamental tasks that must be executed without delay:

- Data processing
- Data storage and query

The primary objective of the data processing action is usually to analyze incoming messages for any abnormal events reported by the IoT device. If any anomalies are detected, appropriate actions need to be taken depending on the severity of the event. These measures can range from sending out SMS alerts to initiating email notifications, ensuring a timely response to any potential issues.

The storage of data and subsequent querying form the backbone of data collection from IoT devices. These actions are not merely rudimentary procedures but serve a crucial role in the overall operation. The data collected is often utilized to maintain a historical record of operational statuses, and it forms the basis for future forecasting. Furthermore, it also aids in the creation of visual representations that provide a clear and concise overview of the system's status.

Now you are all set to proceed with configuring the following two tasks.

Task 1 – abnormal event process

As discussed in the previous chapter, the sensor data in the main ESP32 code is encapsulated in JSON format. You can view the serialized data stream in the local terminal window as shown in the following output:

```
Publishing message: {"timeStamp":1713218277,"deviceMode
l":"DHT11","deviceID":"645ad13bdaec","status":"Normal",
"date":"04-15-2024","time":"14:57:57","timeZone":"-08:0-
0","DST":"Yes","data":{"temp_C":17,"temp_F":63,"humidity":40}}
```

In the AWS IoT Core MQTT test client, the received MQTT message payload will appear as displayed in the following screenshot.

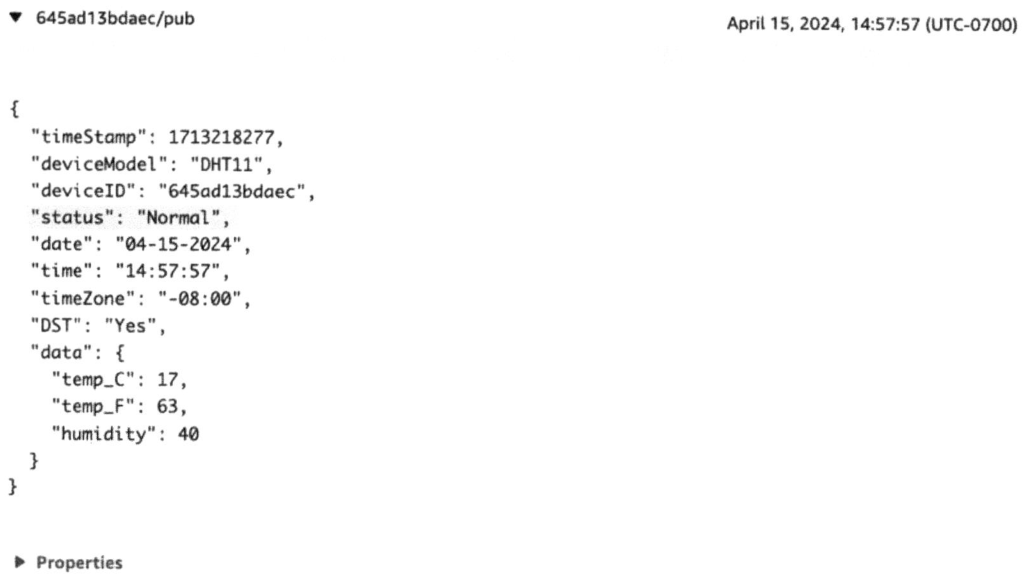

▼ 645ad13bdaec/pub   April 15, 2024, 14:57:57 (UTC-0700)

```json
{
  "timeStamp": 1713218277,
  "deviceModel": "DHT11",
  "deviceID": "645ad13bdaec",
  "status": "Normal",
  "date": "04-15-2024",
  "time": "14:57:57",
  "timeZone": "-08:00",
  "DST": "Yes",
  "data": {
    "temp_C": 17,
    "temp_F": 63,
    "humidity": 40
  }
}
```

▶ Properties

Figure 15.3 – MQTT message shown in AWS IoT Core

You may notice a variable named status in the previous two messages. This variable represents the sensor's status, indicating whether temperature and humidity values fall within, below, or above the normal range. A null value indicates a sensor error. This status is determined by comparing the actual values to a predefined value range in the main code.

In this task, we will instruct AWS IoT Core to forward the payload data to Lambda if the variable named status is not set to Normal.

To understand this service flow, we'll use https://mermaidchart.com, as mentioned in *Chapter 9*, to create a diagram, following these instructions:

```
Create a flow chart to reflect the interaction among the following
AWS services.
```

```
ESP32 sends sensor data in a serialized JSON format to AWS IoT Core.
```

```
AWS IoT Core uses a messaging routing rule to filter the sensor data
status by using an SQL statement.
```

```
If the "status" object is set to "Normal", AWS IoT Core does not
forward this event to Lambda.
```

```
If the "status" object is not set to "Normal", AWS IoT Core forwards
this event to Lambda.
```

Upon receiving non-"Normal" status event, Lambda generates an email content incorporating the sensor data and sends it to SNS (Amazon Simple Notification Service).

Sequentially, SNS sends an email notification to the pre-subscribed user's email address.

The service flow generated as per the preceding instructions is as follows.

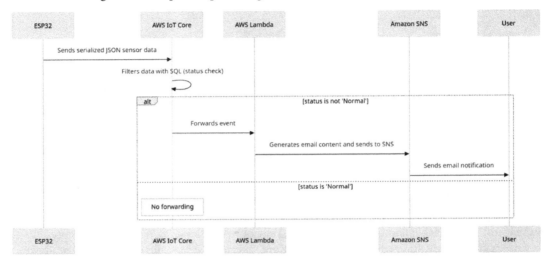

Figure 15.4 – Task 1 service interaction flow

This diagram explicitly describes the task configuration flow, starting from AWS IoT Core to filter received data, creating a forwarding rule to Lambda, and triggering an email notification to SNS.

Configuration steps

The configuration steps for this task, as depicted in *Figure 15.4*, comprise the following phases:

- **Steps 1 to 13; 15 to 18**: Create a message routing rule (triggered by abnormal sensor data through a SQL statement) on AWS IoT Core that points to a new Lambda function
- **Step 14**: Generate a new Lambda function to link to this rule
- **Steps 19 to 23**: Establish an SNS topic and subscription

- **Steps 24 to 31**: Provide the Lambda function with the appropriate role to access SNS
- **Step 32**: Create Python code on this Lambda function to ingest data from AWS IoT Core, produce email content, and send it to SNS to dispatch an email notification to the recipient's address

Let us now look at the steps in detail.

Creating a message routing rule

1. Log in to the AWS console using the regular IAM user we created in the previous steps, that is, **jun.wen**.

Figure 15.5 – Log in as a regular IAM user

2. Search for and click on the **IoT Core** service.

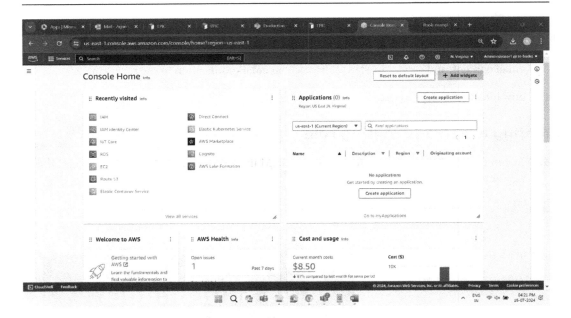

Figure 15.6 – Open the IoT Core console

3.   Click on **Rules**, found under **Message routing** in the left-side navigation menu.

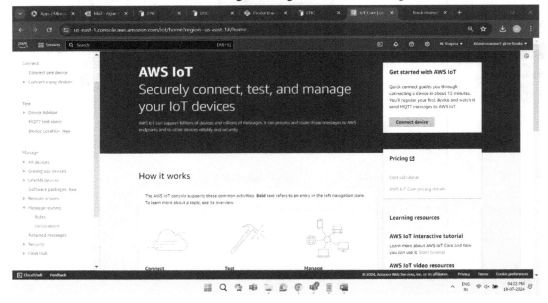

Figure 15.7 – Locate Rules

4. Click on **Create rule** to create a rule from AWS IoT Core to report abnormal events to Lambda.

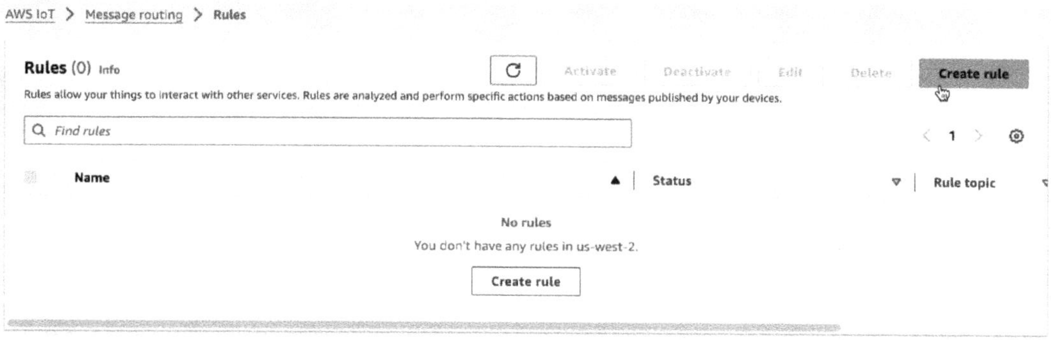

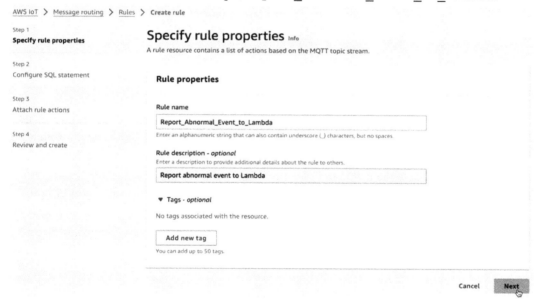

Figure 15.8 – Start to create a rule

5. Give the rule name here, for example, `Report_Abnormal_Event_to_Lambda`.

Figure 15.9 – Name this new rule

6. Create a SQL statement here to report abnormal events to Lambda.

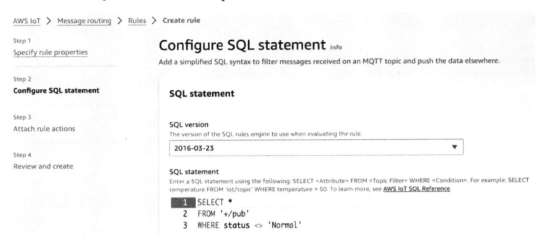

Figure 15.10 – Create a SQL statement

The following is the SQL statement we are going to use:

```
1.  SELECT *
2.  FROM "+/pub"
3.  WHERE status <> 'Normal'
```

The SELECT * and FROM '+/pub' statements represent selecting all objects from a JSON payload and referring to an incoming MQTT publish topic from any ESP32 device with a /pub appendix, where the status is not Normal.

You may recall that in the ESP32 code, we created the publish topic as AWS_IOT_PUBLISH_TOPIC = deviceID + /pub, which means any ESP32 device with an appendix of /pub.

You can also find the AWS IoT SQL reference at https://docs.aws.amazon.com/iot/latest/developerguide/iot-sql-reference.html?icmpid=docs_iot_hp_act.

7. After typing in the SQL statement, navigate to **Rule actions** and click **Choose an action** under **Action 1**.

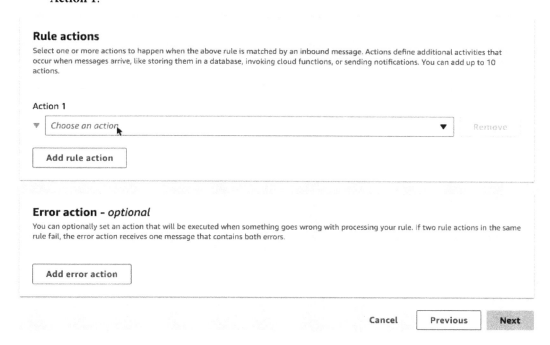

Figure 15.11 – Create an action triggered by this rule

8. Search for and select **Lambda**.

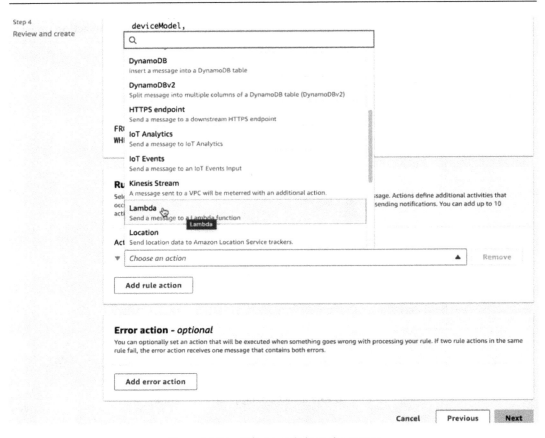

Figure 15.12 – Select Lambda as the action

9.   Click on **Create a Lambda function**, and you will be redirected to the Lambda console.

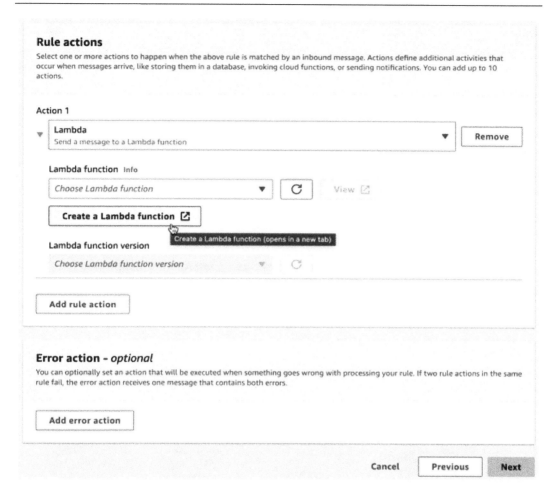

Figure 15.13 – Start to create a Lambda function

10. In the Lambda console, create a Lambda function. Assign a name to the function, such as
Process_Abnormal_Event_from_IoT_Core, and select **Runtime**. AWS Lambda
supports many languages to write functions, such as Python, Java, Dot NET, Ruby, and Node.
js. In this task, we are going to use Python. AWS Lambda functions support Python versions
from 3.8 to 3.12 but AWS highly recommends using version 3.12. Select **Python 3.12**, and then
click on **Create function**.

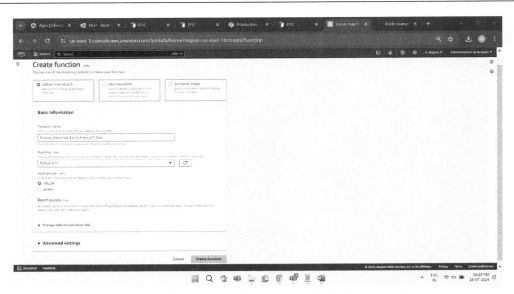

Figure 15.14 – Create a Lambda function with the Python runtime

Creating a Lambda function

A Lambda function named `Process_Abnormal_Event_from_IoT_Core` is successfully created. At this stage, we will not deploy any actual code on it – this will be done in the final step. Since we haven't completed the setup of SNS, we'll return to the previous IoT Core console to continue creating the message routing rules.

Figure 15.15 – Lambda function created (actual code not deployed yet)

Let's begin!

11. In the previous IoT Core console, click **Refresh** under **Action 1**, then click **Lambda function** to load the new Lambda function we just created.

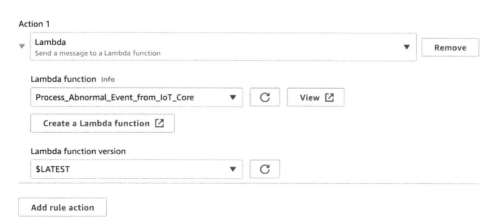

Figure 15.16 – Refresh the Lambda function

12. Select the new Lambda function we just created, and then click **Next**.

Rule actions

Select one or more actions to happen when the above rule is matched by an inbound message. Actions define additional activities that occur when messages arrive, like storing them in a database, invoking cloud functions, or sending notifications. You can add up to 10 actions.

Action 1

Lambda
Send a message to a Lambda function

Remove

Lambda function Info

Process_Abnormal_Event_from_IoT_Core

Create a Lambda function

View

Lambda function version

$LATEST

Add rule action

Figure 15.17 – Select the Lambda function just created

13. Review the configuration information of rules creation and then click **Create**.

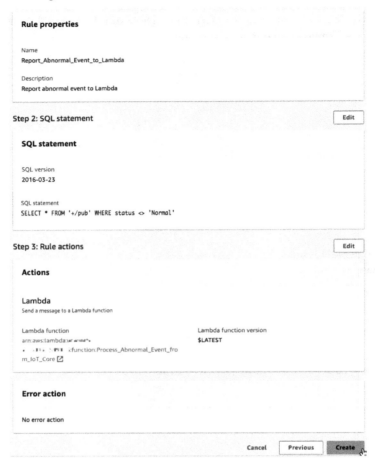

Figure 15.18 – Complete the rule creation

14. Now, you can see that the rule was successfully created.

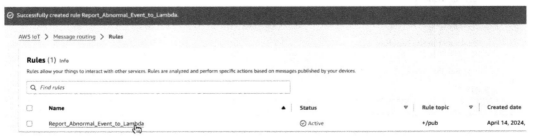

Figure 15.19 – New rule shown under Rules

So far, we have successfully done the following:

- Created a messaging routing rule, allowing IoT Core to report abnormal statuses to the Lambda function

- Created a Lambda function to link IoT Core through the messaging routing rule

Next, we'll finish the remaining configuration before executing the task:

- Create an SNS topic and subscribe an email address to it

- Write Python code on the Lambda function to generate an alert email content under the topic, which is sent to SNS to forward to subscribed email address.

Creating an SNS topic

15. Search for SNS and navigate to the SNS console.

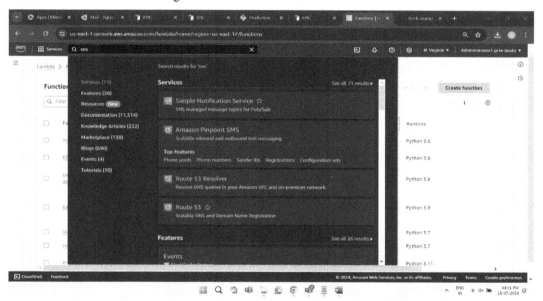

Figure 15.20 – Log in to the SNS console

16. Click on **Topics** in the left-side navigation menu, then click on **Create topic**.

Figure 15.21 – Start to create an SNS topic

17. Under the **Standard** type, enter a topic name, for example, DHT11_Abnormal_Event. The **FIFO** type is a better option for use cases that require message ordering and deduplication. However, in our case, the temperature and humidity abnormal events are standalone, not in any order or correlation, so we select the **Standard** type here.

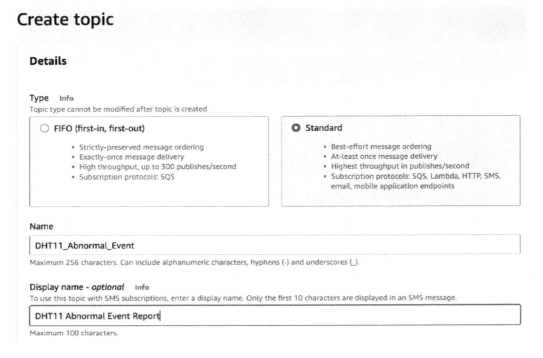

Figure 15.22 – Selecting the Standard type

18. After creating the topic successfully, click on **Create subscription** under this topic.

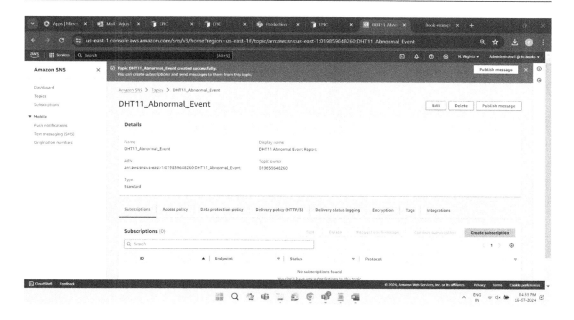

Figure 15.23 – Create subscription

19. Set **Protocol** to **Email**.

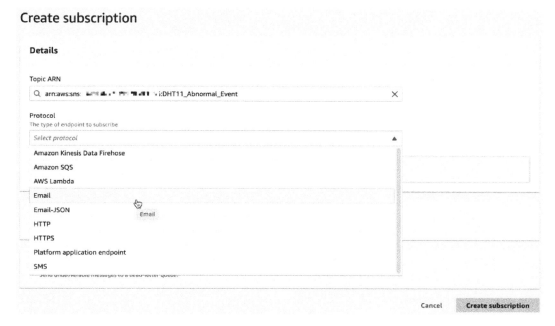

Figure 15.24 – Select the Email protocol

20. In the **Endpoint** section, enter an email address to receive SNS email notifications about any abnormal events. After you create the subscription, a confirmation email will be sent to that recipient address. The recipient must confirm it to begin receiving notification emails.

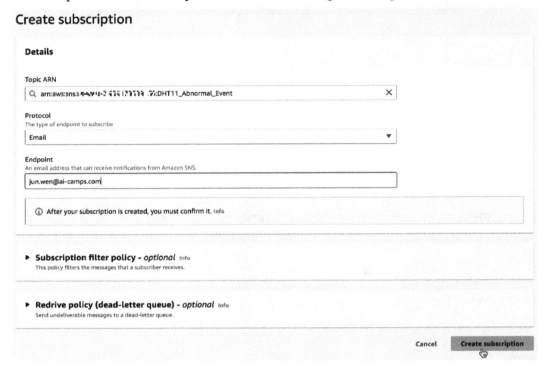

Figure 15.25 – Create a subscription

21. At this point, the SNS email notification is properly set up.

Figure 15.26 – Subscription is confirmed

Now, we have completed the setup of an SNS topic and added an email address to the subscription list. When Lambda sends a notification to SNS with the matching topic, the contents of this notification will be forwarded to this email address.

Now, we have arrived at the final stage of creating Python code on the Lambda function to trigger the notification.

Programming the Lambda function

22. In this final step, we need to program the Lambda function to send abnormal event email notifications to SNS using the previously defined SNS topic.

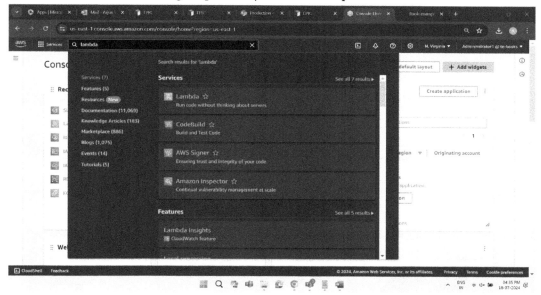

Figure 15.27 – Log in to the Lambda console

23. You will see the previously created Lambda function. Click on its name to edit it.

Figure 15.28 – Edit the Lambda function

24. Before typing code on the Lambda function, we need to grant this function permission to talk to SNS. Click on **Configuration**, then **Permissions**, and then on the **Role** name.

Figure 15.29 – Click the execution role name

25. In this role, click on **Add permissions**.

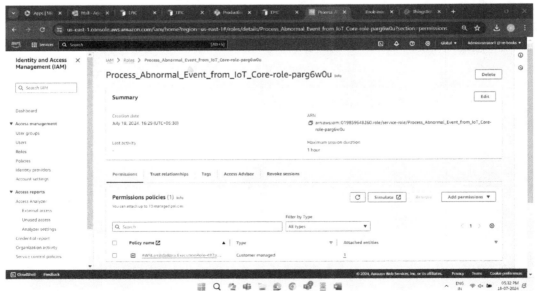

Figure 15.30 – Add permissions

26. Search for and select **AmazonSNSFullAccess**, and then click on **Add permissions**.

Figure 15.31 – Add AmazonSNSFullAccess permission

27. Now, under **Resource summary**, you should see **Amazon SNS**.

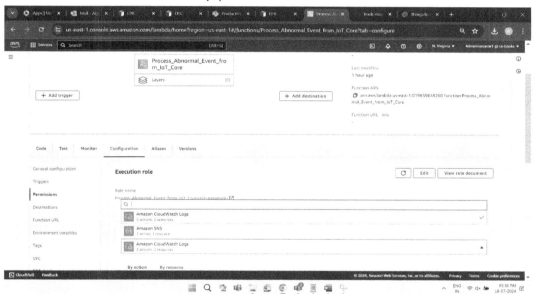

Figure 15.32 – Check whether the Amazon SNS permission is added

28. Now, click on **Code**. We've reached the final step: programming Python code on the Lambda function we previously created.

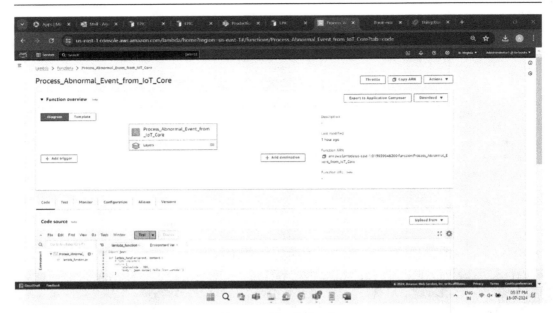

Figure 15.33 – Code tab

29. Since we selected Python 3.12 as the runtime language, we can continue instructing ChatGPT to produce code by our service logic design:

```
Hi, ChatGPT,

Please develop Python 3.12 code on an AWS Lambda function that
meets the following requirements:
```

- AWS IoT Core forwards any abnormal events (where the value of object "status" is not "Normal") to the Lambda function.

- The Lambda function receives JSON format data from IoT Core with the following example structure:

```
1. example:
2. {
3.   "timeStamp": 1713302072,
4.   "deviceModel": "DHT11",
5.   "deviceID": "645ad13bdaec",
6.   "status": "Above Normal",
7.   "date": "04-16-2024",
8.   "time": "14:14:32",
9.   "timeZone": "-08:00",
10.  "DST": "Yes",
11.  "data": {
12.    "temp_C": 19,
```

```
13.     "temp_F": 66,
14.     "humidity": 99
15.   }
16. }
17.
```

Upon receiving the data, the Lambda function should generate an email content in the following format and send it to SNS by the topic of "arn:aws:sns:my_region:my_account_id:DHT11_Abnormal_Event"

```
 1. example:
 2.
 3. Alert: DHT11 Abnormal Event Detected!
 4. Details:
 5. - Status: [value from "status"]
 6. - Device ID: [value from "deviceID"]
 7. - Device Model: [value from "deviceModel"]
 8. - Timestamp: [value from "timeStamp"]
 9. - Date: [value from "date"]
10. - Time: [value from "time"]
11. - TimeZone: [value from "timeZone"]
12. - DST: [value from "DST"]
13. - Temperature (C): [value from "data.temp_C"]
14. - Temperature (F): [value from "data.temp_F"]
15. - Humidity: [value from "data.humidity"]
16.
17. Please take the necessary actions!
18.
```

- You must create two separated functions: format_email_content(received_event) to generate email contents including the event details, and lambda_handler(received_event, context) to send the formatted email contents to the SNS topic.

- Retrieve the SNS topic ARN from environment variables.

You can find this prompt file and a Lambda function Python code example at https://github.com/PacktPublishing/Accelerating-IoT-Development-with-ChatGPT/tree/main/Chapter_15.

30. After obtaining the Python code from ChatGPT, copy and paste it into the **Code** section. To activate the code, click **Deploy**.

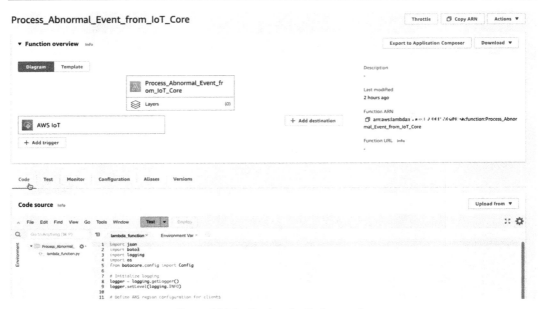

Figure 15.34 – Deploy the Python code

31. Let's conduct a live test by removing the data cable connector on the DHT11 from the ESP32. If all settings are correct, you should receive an email notification, as shown in the following screenshot.

Figure 15.35 – Email notification for abnormal event

32. If the recipient didn't receive the email notification, you can check the logs in the CloudWatch console under the rule name, such as `Process_Abnormal_Event_from_IoT_Core`. Click on the most recent log stream name.

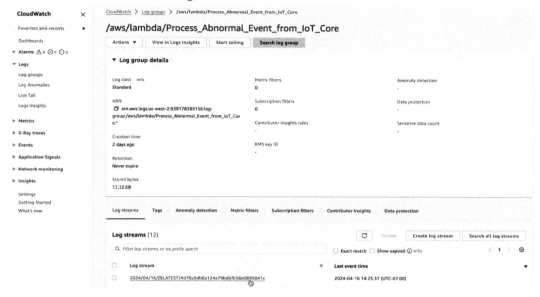

Figure 15.36 – Open the log in the CloudWatch console

33. The following screenshot shows that an abnormal event, **Sensor Error**, was successfully sent as an SNS notification.

Figure 15.37 – Log records

Now, we have successfully set up the abnormal event process task using AWS IoT Core, Lambda, and SNS. When the sensor reports its status at `Sensor Error`, `BelowNormal`, or `AboveNormal`, an alert email notification will be sent out to the recipient.

It is also possible to create an IoT rule to send the abnormal data directly from IoT Core to an SNS topic with original JSON format data, whereas, Lambda can create much more readable format with customized email topic and contents .

Next, we will handle the task of data storage and querying on IoT Analytics, which is quite straightforward.

Task 2 – data storage and querying

In contrast to the abnormal event process, in this task, AWS IoT Core will forward all received data to AWS IoT Analytics, regardless of whether the status is `Normal` or not.

As introduced in *Chapter 5*, AWS IoT Analytics incorporates several sub-services. These include channels, pipelines, data stores, and datasets. Their functions are briefly described here:

- **Channels**: Utilize your channel to collect raw and unprocessed IoT device data from other AWS services
- **Pipelines**: Transform raw data from your channel into valuable information by using activities to filter, enrich, and convert your raw IoT device data
- **Data stores**: Store your processed IoT device data for future data analysis
- **Datasets**: Process query data from IoT devices to deliver automatic, frequent, and current insights about your devices

We'll continue to use `https://mermaidchart.com` to create a diagram using these instructions:

```
Please create a flow chart to reflect the interaction among the
following AWS services.
```

- ESP32 sends sensor data in a serialized JSON format to AWS IoT Core.
- AWS IoT Core uses a messaging routing rule to forward all the received data to AWS IoT Analytics.
- The raw data will arrive at data channels first.
- And then transform data at data pipelines.
- And then store the enriched data at data stores.
- And then provide query to data sets.

The following is the generated service flow as instructed.

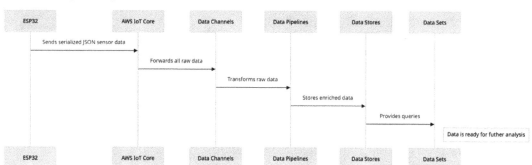

Figure 15.38 – Data storage and query flow

The flow chart illustrates a data storage flow where an ESP32 device sends serialized JSON sensor data to AWS IoT Core, which forwards the raw data to data channels. These channels transmit the data to data pipelines for transformation. The enriched data is then stored in data stores, where it is accessed via queries to create datasets that are ready for further analysis.

The configuration steps for this task, shown in *Figure 15.38*, include the following actions:

- **Steps 1 to 9**: Create IoT Analytics resources
- **Steps 10 to 14**: Create a second message routing rule on AWS IoT Core to forward all sensor data to IoT Analytics
- **Steps 15 to 18**: Run a data query at IoT Analytics

Creating IoT Analytics resources

Let's go through the steps in detail:

1. In the services console, search for and click on **IoT Analytics**.

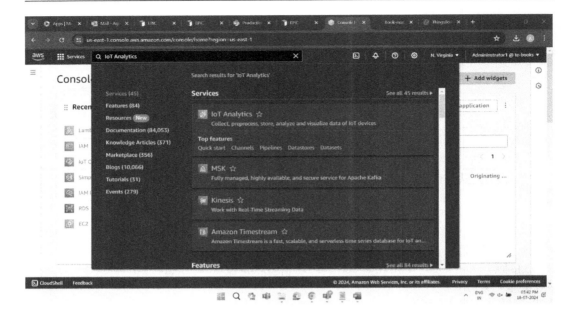

Figure 15.39 – Log in to the IoT Analytics console

2.   Enter a prefix in **Resources prefix**, for instance, dht11.

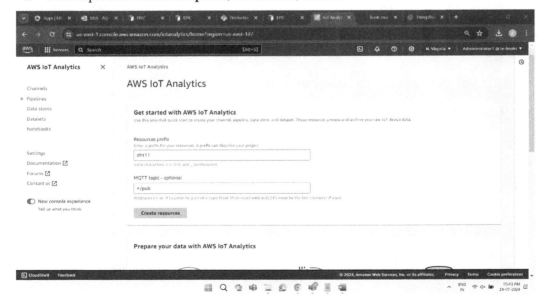

Figure 15.40 – Getting started with IoT Analytics

3. All the IoT Analytics resources will start to be created. There's no need to wait here.

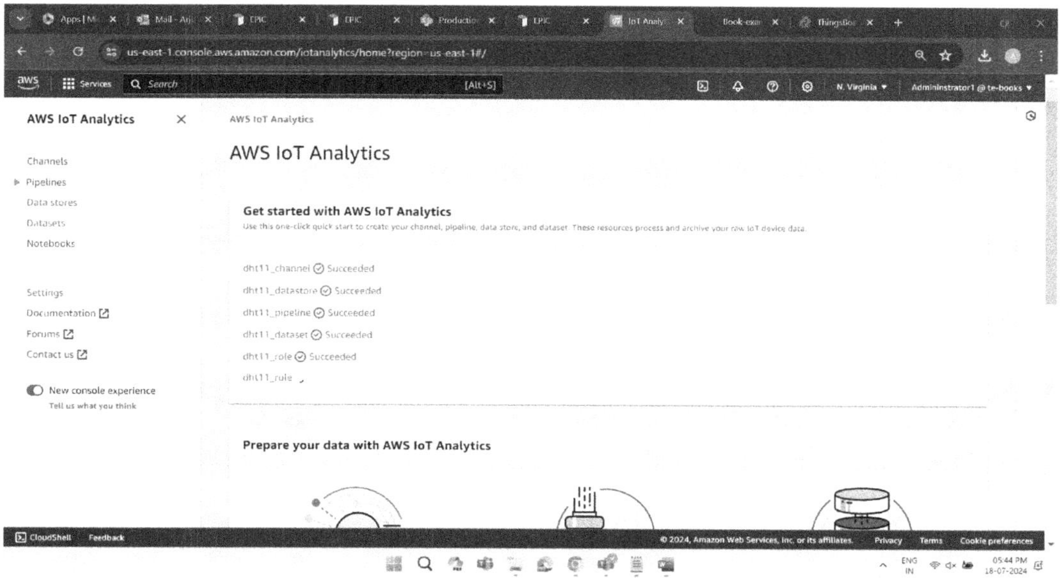

Figure 15.41 – Creation in progress

4. Click on **Channels** in the left-side navigation menu. You'll see a channel named **dht11_channel**, which has already been created.

Figure 15.42 – Check the status of the data channel

5. Click on **Pipelines** in the left-side navigation menu. You will notice that a pipeline named **dht11_pipeline** has already been created.

AWS IoT Analytics > Pipelines

Pipelines (1)

	Name	Created	Last updated
☐	dht11_pipeline	Apr 17, 2024 1:29:03 PM -0700	Apr 17, 2024 1:29:03 PM -0700

Figure 15.43 – Check the status of the data pipeline

6. Click on **Data stores** in the left-side navigation menu. You will find that a data store named **dht11_datastore** has already been created.

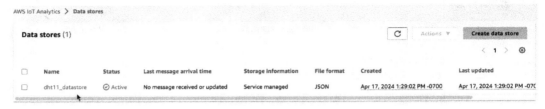

Figure 15.44 – Check the status of data stores

7. Click on **dht11_datastore**, then on **Storage settings**, followed by **Edit**.

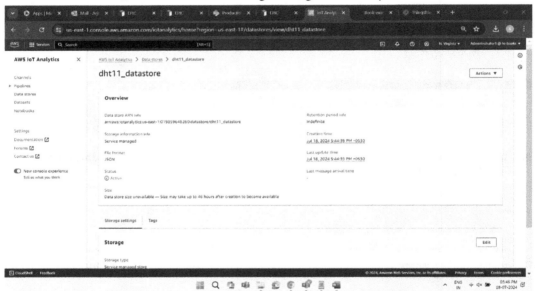

Figure 15.45 – Edit the storage settings

8. You can now set the retention period for your processed data, such as 7 days.

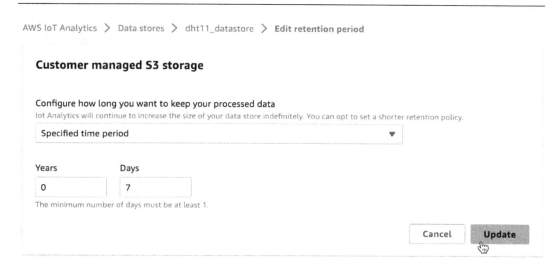

Figure 15.46 – Change the storage time

9. Click on **Datasets** in the left-side navigation menu. You will find a dataset named **dht11_dataset** already created there.

Figure 15.47 – Check the status of the dataset

Now, we have completed the setup of IoT Analytics, including the data channel, data pipeline, data store, and dataset. The subsequent step involves creating another message routing rule in IoT Core, akin to our previous task. This new rule should then be linked to the IoT Analytics resource we created earlier.

Creating a second message routing rule

10. Navigate to the AWS IoT Core console. Under **Message routing**, you will find the first rule we created in the previous task. Now, we are going to create a second rule by clicking on **Create rule**.

Figure 15.48 – Create a new rule in IoT Core

11. Give a rule name here, for example, `Store_All_Events_to_IoT_Analytics`, and click on **Next**.

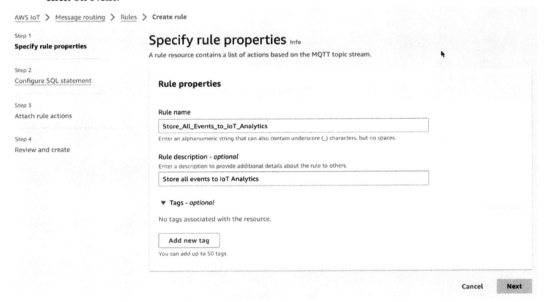

Figure 15.49 – Give a name to this new rule

12. In the **SQL statement** section, add the following statement to forward all the sensor data to IoT Analytics regardless of its status.

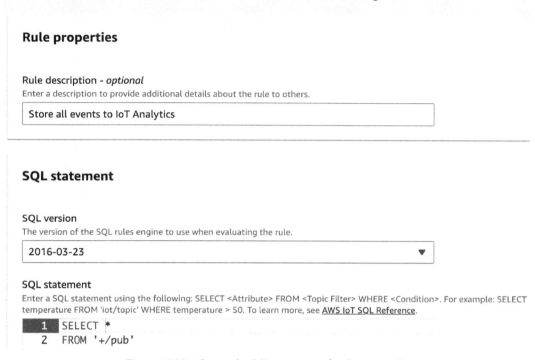

Figure 15.50 – Create the SQL statement for this new rule

13. Under the **Rule actions** section, click on **Action 1**. Search for and select **IoT Analytics**. Under **Channel name**, click on the refresh icon, and then click the **View** button. In the pop-out window, you will find **dht11_channel**, which was created in *step 2*. Under **IAM role**, click on the refresh icon, and then click the **View** button. In the pop-out window, you will find **dht11_role**, which was also created in *step 2*.

Rule actions

Select one or more actions to happen when the above rule is matched by an inbound message. Actions define additional activities that occur when messages arrive, like storing them in a database, invoking cloud functions, or sending notifications. You can add up to 10 actions.

Action 1

▼ | **IoT Analytics**
Send a message to IoT Analytics | ▼ | Remove

Channel name Info

| dht11_channel | ▼ | C | View ☑ |

Create IoT Analytics channel ☑

Batch mode
The payload that contains a JSON array of records will be sent to IoT Analytics via a batch call.

☐ Use batch mode

IAM role
Choose a role to grant AWS IoT access to your endpoint.

| dht11_role | ▼ | C | View ☑ | Create new role |

AWS IoT will automatically create a policy with a prefix of "aws-iot-rule" under your IAM role selected.

Figure 15.51 – Select an action, channel, and role for this rule

14. Now review and create this rule by hitting **Create**.

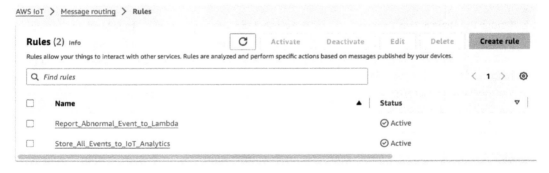

Figure 15.52 – Both rules are shown under Message routing and are now active

Running a data query

At this point, when the sensor data arrives into IoT Core, triggered by the second message routing rule, all the sensor data, regardless of whether it is normal or abnormal, will be ready to be forwardedto IoT Analytics for storage and querying.

Let us run through a data query exercise using the following steps:

15. Click on the dataset name. Under **Details**, click **Edit**. In this step, we will edit the SQL query rule to define the object we want to see from the dataset.

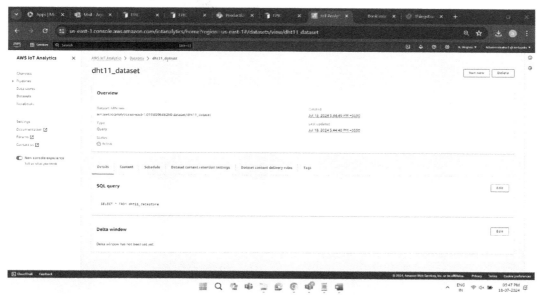

Figure 15.53 – Edit the SQL query statement

For example, if you want to query all the object data in a record and all the records are shown by the timestamp in descending order (the latest record shows at the top), you can edit the current SQL query as in the following screenshot (an example is also provided at https://github.com/PacktPublishing/Accelerating-IoT-Development-with-ChatGPT/blob/main/Chapter_15/dataset_sql_statement).

Figure 15.54 – Update the statement to show records by timestamp in descending order

16. Make sure your ESP32 is operational and successfully connected to AWS IoT Core. Then, go to the **IoT Analytics** console. Under **Datasets**, click on **Run now** to execute a data query. Following that, under **Content**, click the query record name, as shown in the following screenshot.

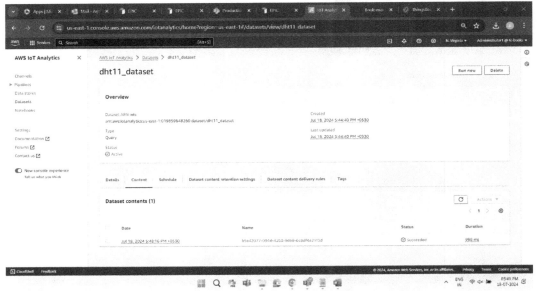

Figure 15.55 – Run data query

17. Finally, you will see all the sensor data stored in the datasets, regardless of status.

Figure 15.56 – Data query result

At this point, we have successfully set up the data storage and query task using AWS IoT Core and IoT Analytics. All sensor data will be stored and ready for analytics queries.

Summary

In this chapter, you learned about the configuration steps to complete two fundamental tasks in a commercial IoT application. The first task is data processing, which involves using AWS Lambda to trigger notifications to SNS to send alert emails about abnormal events. The second task is data storage and querying using AWS IoT Analytics, a crucial aspect of IoT applications that ensures the efficient handling and retrieval of vast amounts of data. These tasks form the backbone of IoT applications and mastering them is key to the successful implementation of any commercial IoT project.

Next, we will explore the final step: executing data analytics and visualization. In the following chapter, we will use the ThingsBoard cloud platform to ingest real-time sensor data from AWS IoT Core to create a dynamic dashboard.

16

Creating a Data Visualization Dashboard on ThingsBoard

In the previous chapter, we successfully processed abnormal events and handled data storage and queries. These steps allowed us to effectively onboard sensor data to the cloud, preparing it for analysis and visualization.

In this chapter, we'll proceed to the final step and go through the following topics:

- Integrating the AWS cloud with ThingsBoard
- Task 1 – Provisioning a ThingsBoard agent with AWS
- Task 2 – Creating the data converter and integrating it with ThingsBoard
- Task 3 – Producing a real-time dashboard with ThingsBoard

By the end of this chapter, you will be equipped with the knowledge to transform your innovation into a commercial-grade solution using this vibrant dashboard.

Technical requirements

In this chapter, in addition to access to your current AWS account, it is assumed that you have already enrolled in a ThingsBoard Cloud account at `https://thingsboard.cloud`.

Integrating the AWS cloud with ThingsBoard

ThingsBoard (`https://thingsboard.io/`) is a renowned open source IoT platform used for data collection, processing, visualization, and device management. It supports device connectivity via industry-standard IoT protocols such as MQTT, CoAP, and HTTP, and can be deployed both on the cloud and on-premises. With its scalable, fault-tolerant, and high-performance features, ThingsBoard ensures your data is always secure and accessible.

ThingsBoard is a user-friendly platform ideal for beginners starting their IoT innovation journey. It allows for quick data ingestion from your device and visualizes the data in a customized real-time dashboard.

To integrate the AWS cloud as an IoT data source, ThingsBoard provides detailed guidance at https://thingsboard.io/docs/user-guide/integrations/aws-iot/.

The integration approach consists of three steps:

1. **Provision a thing in AWS IoT Core**: This *thing* will represent a ThingsBoard agent, which will subscribe to a dedicated topic published by IoT devices. This *thing* will generate its own certificates and attach the policy we created in *Chapter 13*.

2. **Generate a data converter in ThingsBoard**: This converter will be responsible for transforming the received payload from the published messages into the output data you want to display on the final dashboard.

3. **Create an integration in ThingsBoard**: This integration instance will utilize the agent's certificates to access AWS IoT Core, subscribe to the published topic, and incorporate the data converter.

As long as this instance is active, it indicates that the integration is successful and ready to ingest real-time data from the AWS cloud to build the final dashboard at ThingsBoard.

Let us start the configuration guidance with the following tasks.

Task 1 – Provisioning a ThingsBoard agent with AWS

To successfully receive sensor data from AWS IoT Core, it's necessary to have a specific *thing* provisioned in AWS IoT Core, representing ThingsBoard Cloud. This agent is used to subscribe to the MQTT messages published by the target devices.

After creating this specific *thing* in AWS IoT Core, ThingsBoard needs to adopt the certificates that are generated when this agent is created to gain access to AWS IoT Core. This is a critical step as it enables secure and verified communication between the two platforms.

In essence, the agent is more than just a simple conduit for data. It serves as a service integration anchor point, bridging the gap between AWS IoT Core and ThingsBoard Cloud. It ensures the seamless transfer of data from one point to another, all while maintaining the integrity and security of the information being transmitted. Subscribing to MQTT messages from target devices facilitates real-time communication and data exchange, thus effectively linking AWS IoT Core and ThingsBoard Cloud.

Let's begin the task:

1. Log in to the AWS console as the regular user role, navigate to the **IoT Core** service, and click on **Create things**.

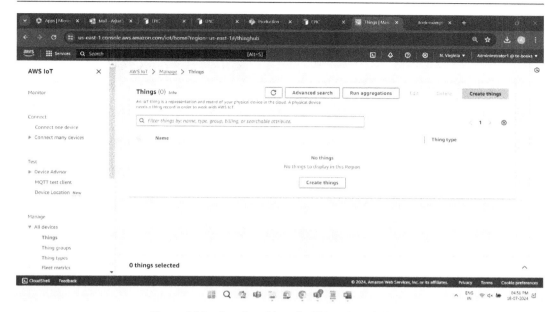

Figure 16.1 – Creating things for ThingsBoard Cloud

2. Create a single *thing*.

AWS IoT > Manage > Things > **Create things**

Create things info

A thing resource is a digital representation of a physical device or logical entity in AWS IoT. Your device or entity needs a thing resource in the registry to use AWS IoT features such as Device Shadows, events, jobs, and device management features.

Number of things to create

⦿ Create single thing
Create a thing resource to register a device. Provision the certificate and policy necessary to allow the device to connect to AWS IoT.

○ **Create many things**
Create a task that creates multiple thing resources to register devices and provision the resources those devices require to connect to AWS IoT.

Cancel   **Next**

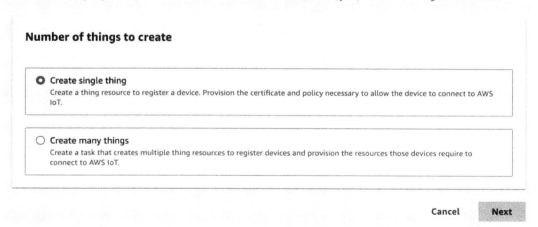

Figure 16.2 – Creating a single thing for ThingsBoard Cloud

3.   Give the *thing* a name here, for example, `ThingsBoard_Agent`.

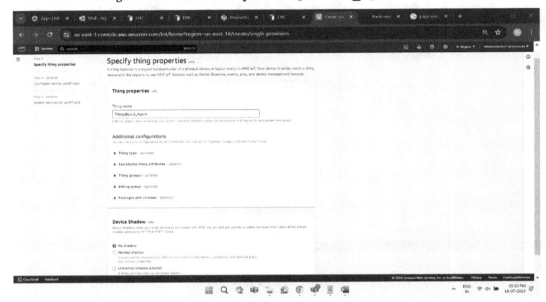

Figure 16.3 – Give the thing a name

4.   Create a new device certificate.

AWS IoT > Manage > Things > Create things > Create single thing

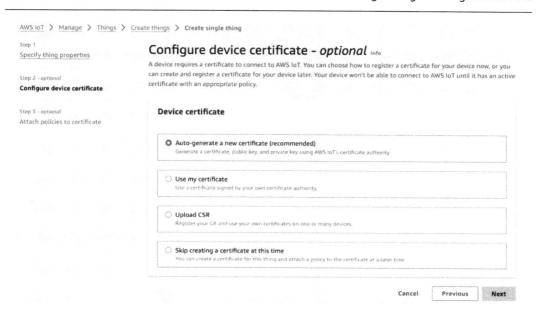

Figure 16.4 – Creating a new certificate for this thing

5.   Attach the policy that we created in *Chapter 13*.

Figure 16.5 – Attaching the policy to this new certificate

Download all the certificates and keys here. Since the AWS IoT integration supported by ThingsBoard Cloud only requests a device certificate, private key file, and Amazon Root CA 1, you just need to import these three files, as shown in *Figure 16.6*.

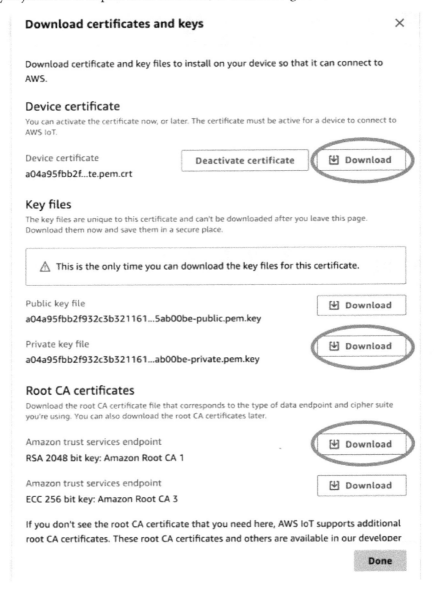

Figure 16.6 – Downloading and saving certificate files

A new *thing* named `ThingsBoard_Agent` has been successfully created in AWS IoT Core.

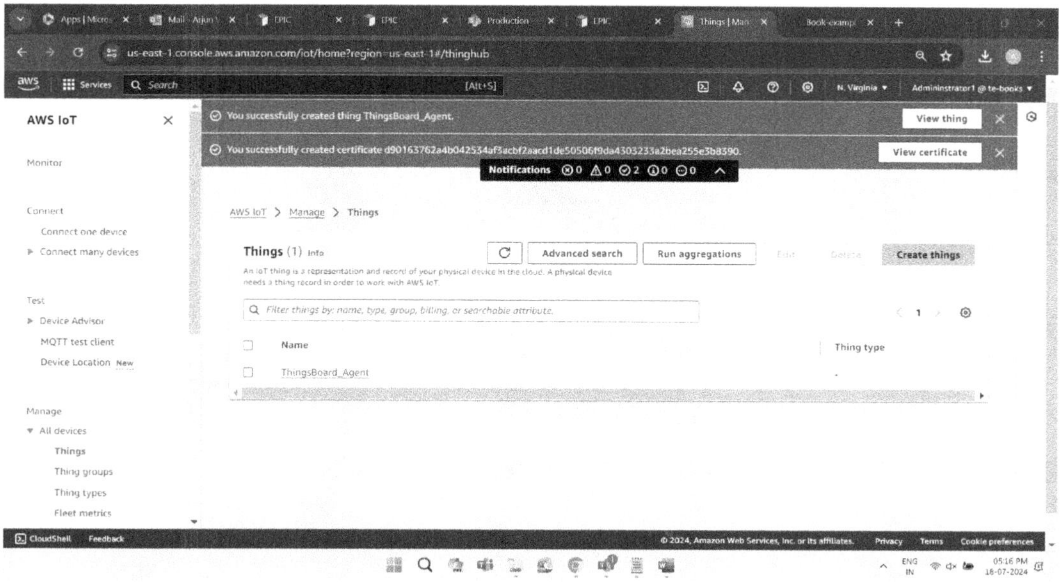

Figure 16.7 – The thing's successful creation page

6. In **Settings**, please note down the **Endpoint** information, such as `your_account_ID-ats-iot.your_region.amazonaws.com`. We will need this information for the integration process in ThingsBoard.

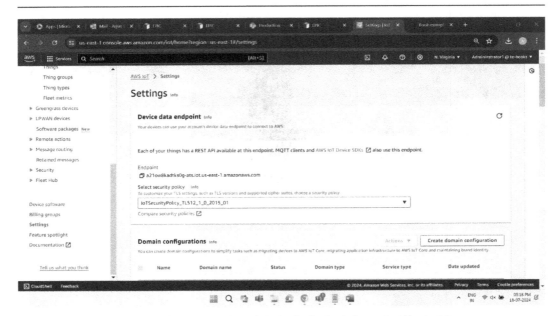

Figure 16.8 – Locating the Endpoint information

At this point, we have successfully created `ThingsBoard_Agent` as a device in AWS IoT Core, downloaded it with its own certificate files, and attached the existing policy. In the next task, we are going to complete the integration task at ThingsBoard.

Task 2 – Creating the data converter and integrating it with ThingsBoard

In addition to receiving published messages from AWS IoT Core, ThingsBoard needs to execute data decoder function to parse the payload of the incoming message and transform it to the format that ThingsBoard uses.

The data converter process in ThingsBoard Cloud involves decoder function creation. ThingsBoard adopts **ThingsBoard Expression Language** (**TBEL**) (`https://thingsboard.io/docs/pe/user-guide/tbel/`) to create a decoder function to facilitate data processing and manipulation. TBEL allows you to write expressions that can parse, transform, and process device data as it is ingested into ThingsBoard.

In the context of a decoder function, TBEL is utilized to interpret incoming JSON, text, or binary (Base64) format data from devices, typically sent over IoT protocols such as MQTT or CoAP. The decoder function then extracts values from the raw device data, transforms them if necessary, and formats them into a structure that ThingsBoard can use for further processing and visualization.

You can find more details at `https://thingsboard.io/docs/user-guide/integrations/`.

For instance, in our project, the JSON payload published from ESP32 to AWS IoT Core is in the following format:

```
1.  {
2.    "timeStamp": 1713817262,
3.    "deviceModel": "DHT11",
4.    "deviceID": "645ad13bdaec",
5.    "status": "Normal",
6.    "date": "04-22-2024",
7.    "time": "13:21:02",
8.    "timeZone": "-08:00",
9.    "DST": "Yes",
10.    "data": {
11.      "temp_C": 22,
12.      "temp_F": 72,
13.      "humidity": 38
14.    }
15.  }
```

For the data visualization dashboard on ThingsBoard Cloud, we might *not* need to incorporate all the objects from this input payload. Instead, we plan to use the following output objects to construct our dashboard:

```
1.  {
2.      "eventType": "Normal",
3.      "deviceName": "645ad13bdaec",
4.      "deviceType": "DHT11",
5.      "telemetry": {
6.          "temp_c": 22,
7.          "temp_f": 72,
8.          "humidity": 38
9.      }
10.  }
```

The following TBEL code example is used for data conversion in our case. It can also be found at https://github.com/PacktPublishing/Accelerating-IoT-Development-with-ChatGPT/tree/main/Chapter_16:

```
1.  // Decode payload to JSON
2.  var received_event = decodeToJson(payload);
3.
4.  // Extract device ID and other telemetry values from JSON
5.  var deviceID = received_event.deviceID;
6.  var eventType = received_event.status;
7.  var deviceType = received_event.deviceModel;
8.  var temp_c = received_event.data.temp_C;
9.  var temp_f = received_event.data.temp_F;
10. var humidity = received_event.data.humidity;
11.
12.
13. // Create telemetry object with extracted values
14. var telemetry = {
15.     temp_c: temp_c,
16.     temp_f: temp_f,
17.     humidity: humidity
18.
19. };
20.
21. // Create result object with device ID and telemetry data
22. var result = {
23.     eventType: eventType,
24.     deviceName: deviceID,
25.     deviceType: deviceType,
26.     telemetry: telemetry
27.
28. };
29.
30. // Helper function to decode JSON
31. function decodeToJson(payload) {
32.     var str = String.fromCharCode.
apply(null, new Uint8Array(payload));
33.     var received_event = JSON.parse(str);
34.     return received_event
35. }
36.
37. return result;
```

Now, let us go through the steps to complete the data converter and integration on ThingsBoard Cloud:

1. Log in to `https://thingsboard.cloud`. Then, click on **Data converters** under **Integrations center** in the left-side navigation bar.

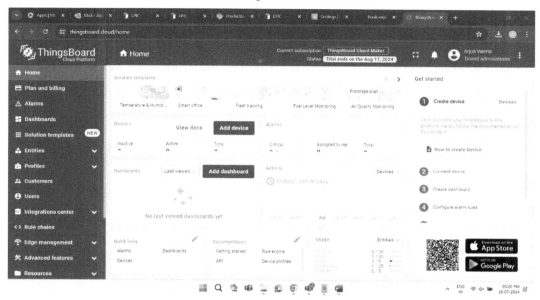

Figure 16.9 – Locating Integrations center in ThingsBoard Cloud

2. Click on + and then on **Create new converter**.

Figure 16.10 – Starting to create a new converter

3. Give a name to this data converter, such as `AWS IoT Uplink Converter`, set **Type** to **Uplink**, and then click on **Test decoder function**.

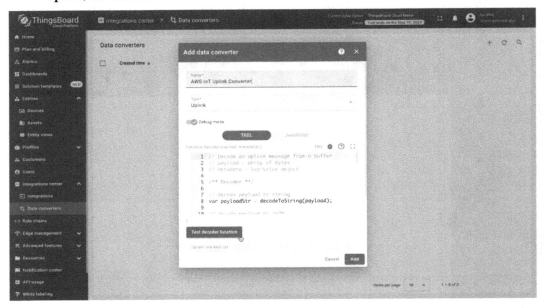

Figure 16.11 – Creating the decoder function

4. In this step, we'll create a decoder function under **TBEL**, as we discussed at the beginning of this section. This decoder function will convert the input payload from AWS IoT Core to our desired output data format.

Figure 16.12 – Testing our decoder function

You can copy and paste the example input into the payload contents. Then, paste the data converter code into the function decoder. Click **Test** to verify the output and save it to move to the next step.

5.   Click on **Add** to add this data converter.

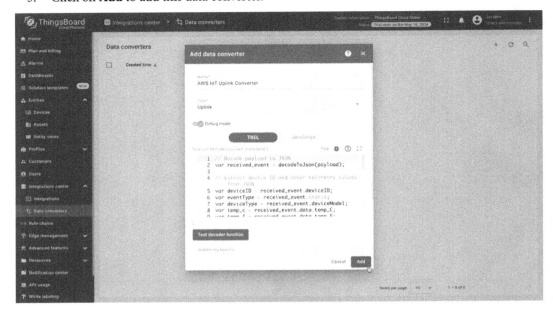

Figure 16.13 – Adding the data converter

6.   Now, this data converter has been successfully created in ThingsBoard.

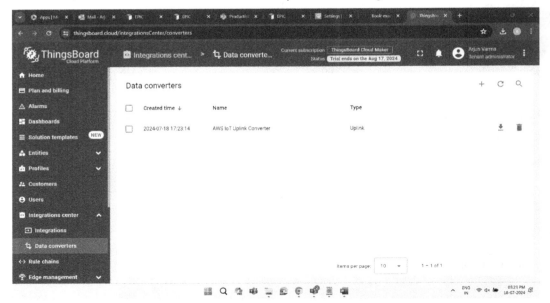

Figure 16.14 – Completing the data converter

7. Now, move to **Integrations** under **Integrations center** to add an integration instance.

Figure 16.15 – Starting to add integration

8. Under **Integration type**, select **AWS IoT**.

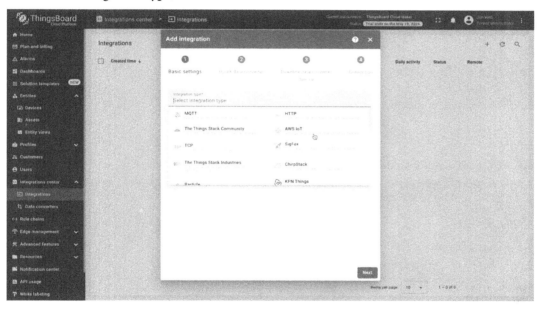

Figure 16.16 – Setting the integration type to AWS IoT

9. Enable the three options and click **Next**.

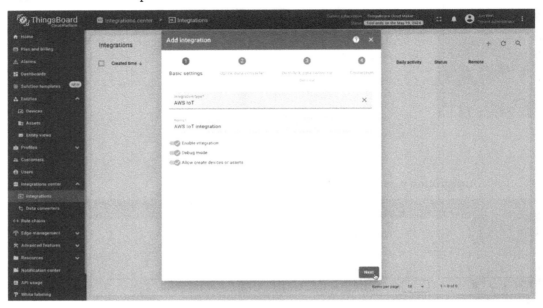

Figure 16.17 – Enabling the options

10. Select the existing uplink data converter we just created and click **Next**.

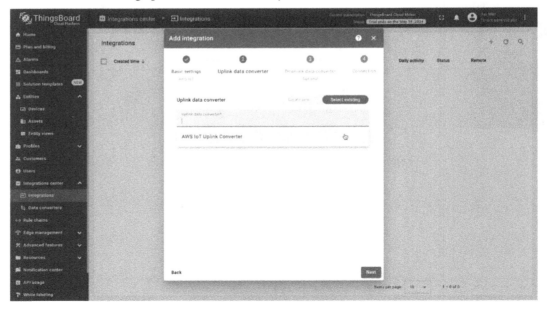

Figure 16.18 – Selecting the data converter created in step 6

11. Skip **Downlink data converter**.

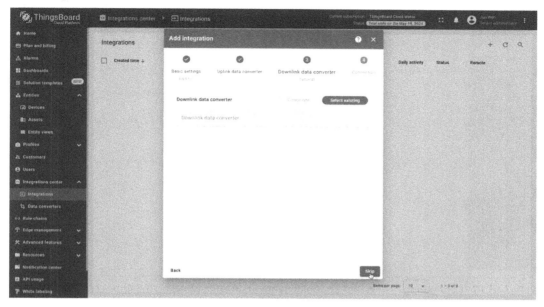

Figure 16.19 – Skipping Downlink data converter

12. Now, enter the AWS IoT **Endpoint** information that we noted in *step 6*, *Figure 16.8* of the previous task. It should look like `xxxxxxxxxxxxxx-ats.iot.your_region.amazonaws.com`. Also, upload the three certificate files downloaded in *step 5* of the previous task.

Figure 16.20 – Configuring the AWS IoT Endpoint information and uploading the three certificate files

13. Finally, under **Topic filters**, enter the topic we created in AWS, +/pub, and click on **Add** to add this integration instance.

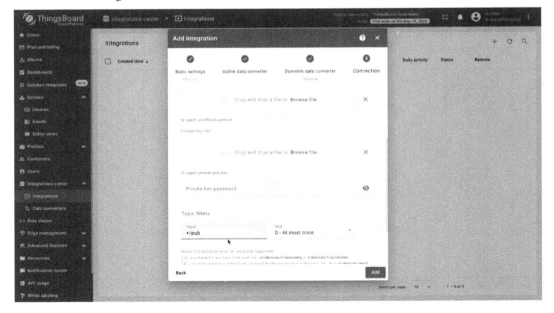

Figure 16.21 – Defining the published topic you want to subscribe

14. You will notice that the integration status is **Pending**.

Figure 16.22 – The integration status is Pending after creation

15. Click the *Refresh* icon in a few seconds, and you will see the status changed to **Active**, which means that the integration was successful.

Figure 16.23 – The integration status changes to Active after refreshing

16. Click on this integration. Under **Event**, you will see the received message, as shown in the following example.

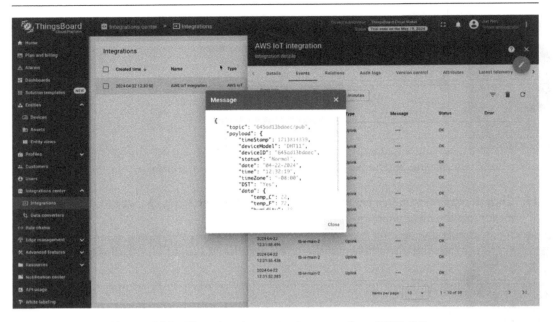

Figure 16.24 – Observing the received message from AWS IoT Core

17. Click on **Data converters**. Under **Event**, you will see the received payload listed under **In** and the output payload under **Out**, as illustrated in the following figure.

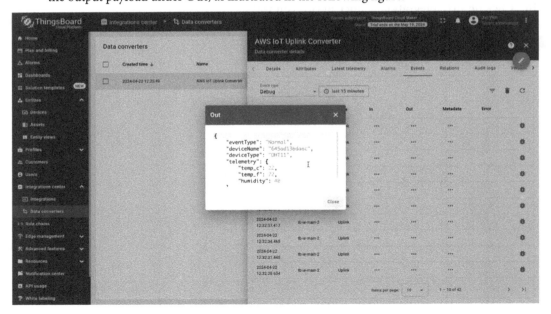

Figure 16.25 – Observing the converted data from the decoder function

18. Click on **Entities**. Under **Device**, you will see that the ESP32 device information has been recorded and is marked as **Active**.

Figure 16.26 – New device showing in Devices in ThingsBoard

19. Click the device, and under **Latest telemetry**, you can see the value read from ESP32.

Figure 16.27 – Observing the converted sensor data under the device

By now, the integration task has been successfully completed! Now, we can move to the final step to produce a real-time dashboard at ThingsBoard.

Task 3 – Producing a real-time dashboard with ThingsBoard

In this task, we are going to use the **Dashboards** feature provided by ThingsBoard.

The *Data Dashboard* feature in ThingsBoard offers a range of tools for creating customizable dashboards. These can include a variety of widgets such as charts, maps, gauges, and tables, designed for real-time and historical data analysis. Users can set up dashboards for dynamic monitoring with real-time data streaming and conduct detailed historical data analysis using advanced data aggregation capabilities such as averages, sums, and counts. Dashboards are interactive, facilitating direct parameter manipulation and thorough data exploration. They can be shared with controlled access for collaboration without compromising security. Furthermore, the responsive design ensures usability across various devices and screen sizes, an essential aspect for effective IoT environment monitoring.

In addition to customizing with the **Dashboards** feature, you can browse **Solution templates** to check whether some templates meet your application requirements.

In this section, we will use the **Dashboards** feature to create our own dashboard:

1.   Under **Dashboards**, create a new dashboard.

Figure 16.28 – Starting to create a new dashboard

2.   Here, you can provide the title and description for the dashboard, such as DHT11 and Temperature and Humidity Data Measurement.

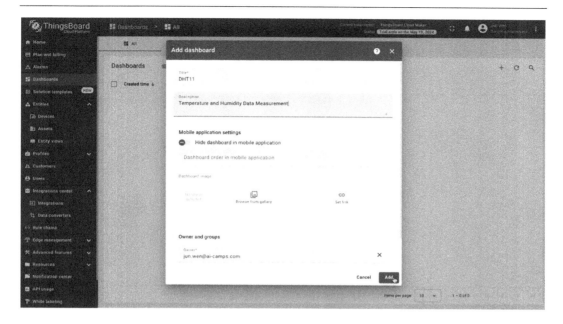

Figure 16.29 – Giving a name to the new dashboard

3.   On the new **DHT11** dashboard, proceed to add a new widget.

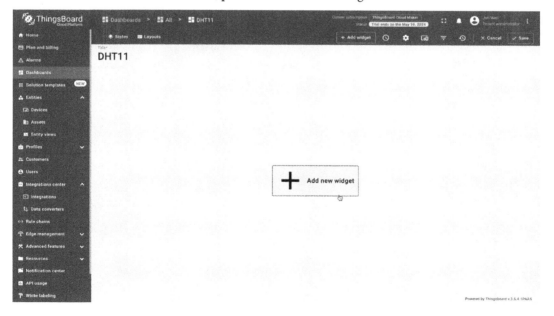

Figure 16.30 – Adding new widgets

4. ThingsBoard offers a variety of widget bundles. Choose one that fits your use case. For our project, we'll start with **Charts**.

Figure 16.31 – Selecting Charts from the widget bundles

5. In the **Charts** section, select **Line chart** from the various available widgets.

Figure 16.32 – Selecting Line chart

6.   Click on **Device** under **Datasource**, where you will find the device ID reported by ESP32. Select it.

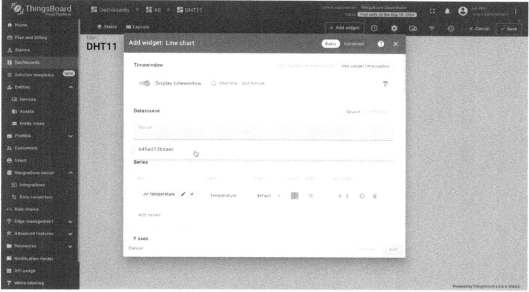

Figure 16.33 – Selecting the target device to display its data

7.   Go to **Series | Key | temperature**, and click the *Edit* icon to edit this key.

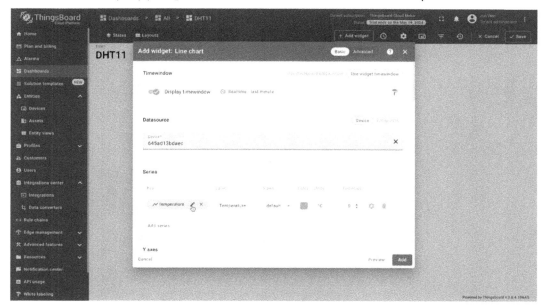

Figure 16.34 – Starting to configure data entry (key)

8.   Click **x** on the right of the **Key** value to load the current value.

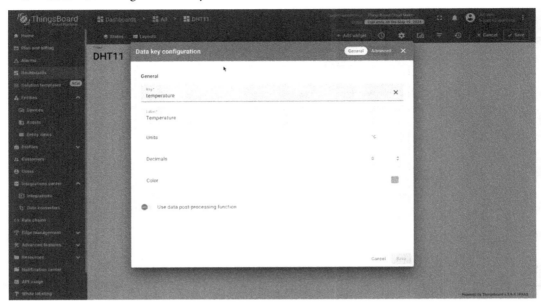

Figure 16.35 – Customizing the data entry format, unit, and appearance color

Now you will see the actual data entry (*key*), such as **temp_c**, **temp_f**, and **humidity**.

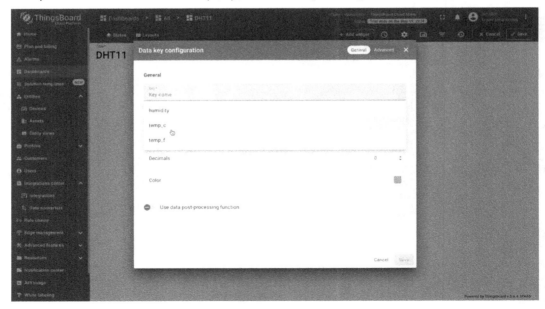

Figure 16.36 – Selecting what data entry you want to show

For example, select **temp_c** and click **Add**. You will see the dynamic line chart of the **temp_c** data in the dashboard.

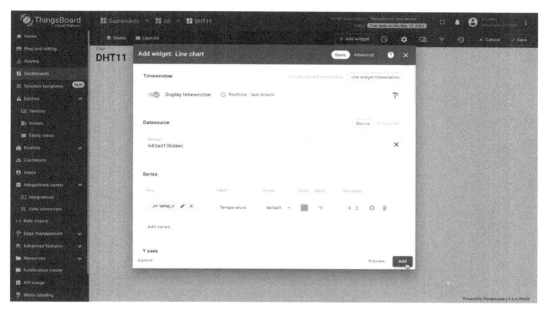

Figure 16.37 – Selecting the data entry of temp_c

By repeating the preceding steps, you can display three-line charts for **temp_c**, **temp_f**, and **humidity**. You can also add more widgets to the dashboard as needed.

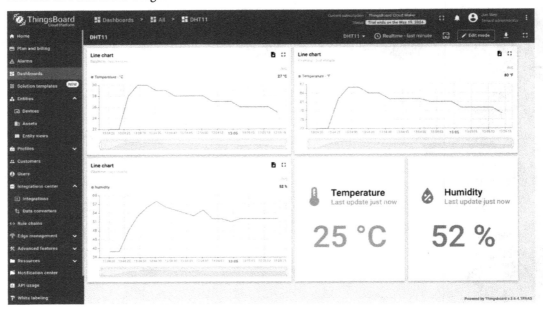

Figure 16.38 – The dashboard appearance

You have now completed the task of creating a data visualization dashboard of a temperature and humidity sensor on ThingsBoard! You can now finally show the real-time sensor data on this dashboard, and you can change the environmental temperature and humidity conditions around your ESP32, and then observe the data chart change on your dashboard accordingly!

Summary

Reflecting on our journey from *Chapter 11* onward, with the assistance of ChatGPT, we began programming the first line of C++ code on ESP32. We collected DHT11 sensor data, connected it to our home Wi-Fi network, and accessed and published the sensor data to the AWS cloud. We created multiple tasks in the AWS cloud to handle abnormal events and stored the sensor data for future queries. Finally, we completed the integration of ThingsBoard to ingest data from the AWS cloud and created a real-time dashboard to display the dynamic data reported by the ESP32!

By now, you should be able to observe your sensor's real-time data on the dashboard! It truly is a long journey for such a tiny packet of data, with several bytes flying from an ESP32 board on your table, crossing the Internet, arriving at AWS, and finally showing their meaningful value on a commercial-grade dashboard. Among these elements in this journey, the most challenging part is programming C++ code on the ESP32 to capture the sensor data and coordinate sensors, LEDs, and buzzers to work together. Thanks to the smart coding skills of ChatGPT and other AI tools, you don't have to worry too much. By using effective prompting frameworks and skills, you can instruct ChatGPT to create C++ code based on your innovative ideas!

Looking ahead, the integration of AI tools in IoT development is poised to become even more sophisticated. As AI continues to evolve, it will bring new capabilities that simplify complex coding tasks and enhance the overall development process. For instance, future advancements may include more intuitive AI-driven debugging tools, advanced code optimization techniques, and seamless integration with various IoT platforms.

Remember, the journey of IoT development is continuous and ever-evolving. Embrace the challenges, stay curious, and keep experimenting with new tools and techniques. With dedication and the right resources, you'll be well-equipped to bring your innovative IoT ideas to life and make a meaningful impact in the world of connected devices.

Index

packtpub.com

Subscribe to our online digital library for full access to over 7,000 books and videos, as well as industry leading tools to help you plan your personal development and advance your career. For more information, please visit our website.

Why subscribe?

- Spend less time learning and more time coding with practical eBooks and Videos from over 4,000 industry professionals

- Improve your learning with Skill Plans built especially for you

- Get a free eBook or video every month

- Fully searchable for easy access to vital information

- Copy and paste, print, and bookmark content

Did you know that Packt offers eBook versions of every book published, with PDF and ePub files available? You can upgrade to the eBook version at packtpub.com and as a print book customer, you are entitled to a discount on the eBook copy. Get in touch with us at customercare@packtpub.com for more details.

At www.packtpub.com, you can also read a collection of free technical articles, sign up for a range of free newsletters, and receive exclusive discounts and offers on Packt books and eBooks.

Other Books You May Enjoy

If you enjoyed this book, you may be interested in these other books by Packt:

Internet of Things Programming Projects

Colin Dow

ISBN: 978-1-83508-295-9

- Integrate web services into projects for real-time data display and analysis
- Integrate sensors, motors, and displays to build smart IoT devices
- Build a weather indicator using servo motors and LEDs
- Create an autonomous IoT robot car capable of performing tasks
- Develop a home security system with real-time alerts and SMS notifications
- Explore LoRa and LoRaWAN for remote environmental monitoring

Artificial Intelligence for Robotics

Francis X. Govers III

ISBN: 978-1-80512-959-2

- Get started with robotics and AI essentials
- Understand path planning, decision trees, and search algorithms to enhance your robot
- Explore object recognition using neural networks and supervised learning techniques
- Employ genetic algorithms to enable your robot arm to manipulate objects
- Teach your robot to listen using Natural Language Processing through an expert system
- Program your robot in how to avoid obstacles and retrieve objects with machine learning and computer vision
- Apply simulation techniques to give your robot an artificial personality

Packt is searching for authors like you

If you're interested in becoming an author for Packt, please visit authors.packtpub.com and apply today. We have worked with thousands of developers and tech professionals, just like you, to help them share their insight with the global tech community. You can make a general application, apply for a specific hot topic that we are recruiting an author for, or submit your own idea.

Share your thoughts

Now you've finished *Accelerating IoT Development with ChatGPT*, we'd love to hear your thoughts! Scan the QR code below to go straight to the Amazon review page for this book and share your feedback or leave a review on the site that you purchased it from.

https://packt.link/r/183546162X

Your review is important to us and the tech community and will help us make sure we're delivering excellent quality content.

Download a free PDF copy of this book

Thanks for purchasing this book!

Do you like to read on the go but are unable to carry your print books everywhere?

Is your eBook purchase not compatible with the device of your choice?

Don't worry, now with every Packt book you get a DRM-free PDF version of that book at no cost.

Read anywhere, any place, on any device. Search, copy, and paste code from your favorite technical books directly into your application.

The perks don't stop there, you can get exclusive access to discounts, newsletters, and great free content in your inbox daily

Follow these simple steps to get the benefits:

1. Scan the QR code or visit the link below

https://packt.link/free-ebook/978-1-83546-162-4

2. Submit your proof of purchase
3. That's it! We'll send your free PDF and other benefits to your email directly

www.ingramcontent.com/pod-product-compliance
Lightning Source LLC
Chambersburg PA
CBHW080613060326
40690CB00021B/4686

* 9 7 8 1 8 3 5 4 6 1 6 2 4 *